by Steven Holzner, PhD
Author of *Quantum Physics For Dummies*

WILEY
Wiley Publishing, Inc.

Quantum Physics Workbook For Dummies®

Published by
Wiley Publishing, Inc.
111 River St.
Hoboken, NJ 07030-5774
www.wiley.com

Published by Wiley Publishing, Inc., Indianapolis, Indiana

Published simultaneously in Canada

For general information on our other products and services, please contact our Customer Care Department within the U.S. at 877-762-2974, outside the U.S. at 317-572-3993, or fax 317-572-4002.

For technical support, please visit www.wiley.com/techsupport.

Wiley also publishes its books in a variety of electronic formats. Some content that appears in print may not be available in electronic books.

Library of Congress Control Number: 2009939359

ISBN: 978-0-470-52589-0

Manufactured in the United States of America

10 9 8 7 6 5 4 3 2 1

About the Author

Steven Holzner is the award-winning writer of many books, including *Physics For Dummies, Differential Equations For Dummies, Quantum Physics For Dummies,* and many others. He graduated from MIT and got his PhD at Cornell University. He's been in the faculty of both MIT and Cornell.

Dedication

To Nancy, of course.

Author's Acknowledgments

Thanks to everyone at Wiley who helped make this book possible. A big hearty thanks to Tracy Boggier, Acquisitions Editor; Chad Sievers, Project Editor; Danielle Voirol, Senior Copy Editor; Kristie Rees, Project Coordinator; Dan Funch Wohns, Technical Editor; and anyone else I may have failed to mention.

Publisher's Acknowledgments

We're proud of this book; please send us your comments at http://dummies.custhelp.com. For other comments, please contact our Customer Care Department within the U.S. at 877-762-2974, outside the U.S. at 317-572-3993, or fax 317-572-4002.

Some of the people who helped bring this book to market include the following:

Acquisitions, Editorial, and Media Development

Project Editor: Chad R. Sievers

Acquisitions Editor: Tracy Boggier

Senior Copy Editor: Danielle Voirol

Assistant Editor: Erin Calligan Mooney

Editorial Program Coordinator: Joe Niesen

Technical Editors: Dan Funch Wohns, Gang Xu

Editorial Manager: Michelle Hacker

Editorial Assistant: Jennette ElNaggar

Cover Photos: © Kevin Fleming/CORBIS

Cartoons: Rich Tennant (www.the5thwave.com)

Composition Services

Project Coordinator: Kristie Rees

Layout and Graphics: Carrie A. Cesavice, Mark Pinto, Melissa K. Smith

Proofreaders: Laura Albert, Melissa D. Buddendeck

Indexer: BIM Indexing & Proofreading Services

Publishing and Editorial for Consumer Dummies

Diane Graves Steele, Vice President and Publisher, Consumer Dummies

Kristin Ferguson-Wagstaffe, Product Development Director, Consumer Dummies

Ensley Eikenburg, Associate Publisher, Travel

Kelly Regan, Editorial Director, Travel

Publishing for Technology Dummies

Andy Cummings, Vice President and Publisher, Dummies Technology/General User

Composition Services

Debbie Stailey, Director of Composition Services

Contents at a Glance

Table of Contents

Introduction

When you make the leap from classical physics to the small, quantum world, you enter the realm of probability. Quantum physics is an exciting field with lots of impressive results if you know your way around — and this workbook is designed to make sure you do know your way around.

I designed this workbook to be your guided tour through the thicket of quantum physics problem-solving. Quantum physics includes more math than you can shake a stick at, and this workbook helps you become proficient at it.

About This Book

Quantum physics, the study of the very small world, is actually a very big topic. To cover those topics, quantum physics is broken up into many different areas — harmonic oscillators, angular momentum, scattered particles, and more. I provide a good overview of those topics in this workbook, which maps to a college course.

For each topic, you find a short introduction and an example problem; then I set you loose on some practice problems, which you can solve in the white space provided. At the end of the chapter, you find the answers and detailed explanations that tell you how to get those answers.

You can page through this book as you like instead of having to read it from beginning to end — just jump in and start on your topic of choice. If you need to know concepts that I've introduced elsewhere in the book to solve a problem, just follow the cross-references.

Conventions Used in This Book

Here are some conventions I follow to make this book easier to follow:

- The answers to problems, the action part of numbered steps, and vectors appear in **bold.**
- I write new terms in *italics* and then define them. Variables also appear in italics.
- Web addresses appear in `monofont`.

Foolish Assumptions

Here's what I assume about you, my dear reader:

- **You've had some exposure to quantum physics, perhaps in a class.** You now want just enough explanation to help you solve problems and sharpen your skills. If you want a more in-depth discussion on how all these quantum physics concepts work, you may want to pick up the companion book, *Quantum Physics For Dummies* (Wiley). You don't have to be a whiz at quantum physics, just have a glancing familiarity.
- **You're willing to invest some time and effort in doing these practice problems.** If you're taking a class in the subject and are using this workbook as a companion to the course to help you put the pieces together, that's perfect.
- **You know some calculus.** In particular, you should be able to do differentiation and integration and work with differential equations. If you need a refresher, I suggest you check out *Differential Equations For Dummies* (Wiley).

How This Book Is Organized

I divide this workbook into five parts. Each part is broken down into chapters discussing a key topic in quantum physics. Here's an overview of what I cover.

Part I: Getting Started with Quantum Physics

This part covers the basics. You get started with state vectors and with the entire power of quantum physics. You also see how to work with free particles, with particles bound in square wells, and with harmonic oscillators here.

Part II: Round and Round with Angular Momentum and Spin

Quantum physics lets you work with the micro world in terms of the angular momentum of particles as well as the spin of electrons. Many famous experiments — such as the Stern-Gerlach experiment, in which beams of particles split in magnetic fields — are understandable only in terms of quantum physics. You see how to handle problems that deal with these topics right here.

Part III: Quantum Physics in Three Dimensions

Up to this point, the quantum physics problems you solve all take place in one dimension. But the world is a three-dimensional kind of place. This part rectifies that by taking quantum physics to three dimensions, where square wells become cubic wells and so on. You also take a look at the two main coordinate systems used for three-dimensional work: rectangular and spherical coordinates. You work with the hydrogen atom as well.

Part IV: Acting on Impulse — Impacts in Quantum Physics

This part is on perturbation theory and scattering. Perturbation theory is all about giving systems a little shove and seeing what happens — like applying an electric field to particles in harmonic oscillation. Scattering theory has to do with smashing one particle against another and predicting what's going to happen. You see some good collisions here.

Part V: The Part of Tens

The Part of Tens is a common element of all *For Dummies* books. In this part, you see ten tips for problem-solving, a discussion of quantum physics's ten greatest solved problems, and ten ways to avoid common errors when doing the math.

Icons Used in This Book

You find a few icons in this book, and here's what they mean:

This icon points out example problems that show the techniques for solving a problem before you dive into the practice problems.

This icon gives you extra help (including shortcuts and strategies) when solving a problem.

This icon marks something to remember, such as a law of physics or a particularly juicy equation.

Where to Go from Here

If you're ready, you can do the following:

- **Jump right into the material in Chapter 1.** You don't have to start there, though; you can jump in anywhere you like. I wrote this book to allow you to take a stab at any chapter that piques your interest. However, if you need a touchup on the foundations of quantum physics, Chapter 1 is where all the action starts.
- **Head to the table of contents or index.** Search for a topic that interests you and start practicing problems. (***Note:*** I do suggest that you don't choose the answer key as your first "topic of interest" — looking up the solutions before attempting the problems kind of defeats the purpose of a workbook! I promise you're not being graded here, so just relax and try to understand the processes.)

- **Check out *Quantum Physics For Dummies.*** My companion book provides a more comprehensive discussion. With both books by your side, you can further strengthen your knowledge of quantum physics.
- **Go on vacation.** After reading about quantum physics, you may be ready for a relaxing trip to a beach where you can sip fruity cocktails, be waited on hand and foot, and read some light fiction on parallel universes. Or maybe you can visit Fermilab (the Fermi National Accelerator Laboratory), west of Chicago, to tour the magnet factory and just hang out with their herd of bison for a while.

Part I
Getting Started with Quantum Physics

The 5th Wave — By Rich Tennant

In this part . . .

This part gets you started in solving problems in quantum physics. Here, you find an introduction to the conventions and principles necessary to solve quantum physics problems. This part is where you see one of quantum physics's most powerful topics: solving the energy levels and wave functions for particles trapped in various bound states. You also see particles in harmonic oscillation. Quantum physicists are experts at handling those kinds of situations.

Chapter 1

The Basics of Quantum Physics: Introducing State Vectors

In This Chapter

- Creating state vectors
- Using quantum physics operators
- Finding expectation values for operators
- Simplifying operations with eigenvalues and eigenvectors

If you want to hang out with the cool quantum physics crowd, you have to speak the lingo. And in this field, that's the language of mathematics. Quantum physics often involves representing probabilities in matrices, but when the matrix math becomes unwieldy, you can translate those matrices into the bra and ket notation and perform a whole slew of operations.

This chapter gets you started with the basic ideas behind quantum physics, such as the state vector, which is what you use to describe a multistate system. I also cover using operators, making predictions, understanding properties such as commutation, and simplifying problems by using eigenvectors. Here you can also find several problems to help you become more acquainted with these concepts.

Describing the States of a System

The beginnings of quantum physics include explaining what a system's *states* can be (such as whether a particle's spin is up or down, or what orbital a hydrogen atom's electron is in). The word *quantum* refers to the fact that the states are *discrete* — that is, no state is a mix of any other states. A quantum number or a set of quantum numbers specifies a particular state. If you want to break quantum physics down to its most basic form, you can say that it's all about working with multistate systems.

Don't let the terminology scare you (which can be a constant struggle in quantum physics). A *multistate system* is just a system that can exist in multiple states; in other words, it has different energy levels. For example, a pair of dice is a multistate system. When you roll a pair of dice, you can get a sum of 2, 3, 5, all the way up to 12. Each one of those values represents a different state of the pair of dice.

Quantum physics likes to spell everything out, so it approaches the two dice by asking how many ways they could be in the various states. For example, you have only one way to roll a 2 with two dice, but you have six ways to roll a total of 7. So if the relative probability of rolling a 2 is one, the relative probability of rolling a 7 is six.

With a little thought, you can add up all the ways to get a 2, a 3, and so on like this:

Sum of the Dice	***Relative Probability of Getting That Sum***
2	1
3	2
4	3
5	4
6	5
7	6
8	5
9	4
10	3
11	2
12	1

In this case, you can say that the total of the two dice is the quantum number and that each quantum number represents a different state. Each system can be represented by a *state vector* — a one-dimensional matrix — that indicates the relative probability amplitude of being in each state. Here's how to set one up:

1. **Write down the relative probability of each state and put it in vector form.**

   You now have a one-column matrix listing the probabilities (though you can instead use a one-row matrix).

2. **Take the square root of each number to get the probability amplitude.**

   State vectors record not the actual probabilities but rather the probability *amplitude,* which is the square root of the probability. That's because when you find probabilities using quantum physics, you multiply two state vectors together (sometimes with an *operator* — a mathematical construct that returns a value when you apply it to a state vector).

3. **Normalize the state vector.**

   Because the total probability that the system is in one of the allowed states is 1, the square of a state vector has to add up to 1. To square a state vector, you multiply every element by itself and then add all the squared terms (it's just like matrix multiplication). However, at this point, squaring each term in the state vector and adding them all usually doesn't give you 1, so you have to *normalize* the state vector by dividing each term by the square root of the sum of the squares.

4. **Set the vector equal to $|\psi\rangle$.**

   Because you may be dealing with a system that has thousands of states, you usually abbreviate the state vector as a Greek letter, using notation like this: $|\psi\rangle$ (or $\langle\psi|$ if you used a row vector). You see why this notation is useful in the next section.

Check out the following example problem and practice problems, which can help clarify any other questions you may have.

Q. What's the state vector for the various possible states of a pair of dice?

A. $|\psi\rangle =$

$$\begin{vmatrix} \frac{1}{6} \\ \frac{2^{1/2}}{6} \\ \frac{3^{1/2}}{6} \\ \frac{2}{6} \\ \frac{5^{1/2}}{6} \\ \frac{6^{1/2}}{6} \\ \frac{5^{1/2}}{6} \\ \frac{2}{6} \\ \frac{3^{1/2}}{6} \\ \frac{2^{1/2}}{6} \\ \frac{1}{6} \end{vmatrix}$$

Start by creating a vector that holds the relative probability of each state — that is, the first value holds the relative probability (the number of states) that the total of the two dice is 2, the next item down holds the relative probability that the total of the two dice is 3, and so on. That looks like this:

$$\begin{vmatrix} 1 \\ 2 \\ 3 \\ 4 \\ 5 \\ 6 \\ 5 \\ 4 \\ 3 \\ 2 \\ 1 \end{vmatrix}$$

Convert this vector to probability amplitudes by taking the square root of each entry like this:

$$\begin{vmatrix} 1 \\ 2^{1/2} \\ 3^{1/2} \\ 2 \\ 5^{1/2} \\ 6^{1/2} \\ 5^{1/2} \\ 2 \\ 3^{1/2} \\ 2^{1/2} \\ 1 \end{vmatrix}$$

When you square the state vector, the square has to add up to 1; that is, the dice must show a 2, 3, 4, 5, 6, 7, 8, 9, 10, 11, or 12. However, squaring each term in this state vector and adding them all up gives you 36, not 1, so you have to normalize the state vector by dividing each term by the square root of 36, or 6, to make sure that you get 1 when you square the state vector. That means the state vector looks like this:

$$\begin{vmatrix} \frac{1}{6} \\ \frac{2^{1/2}}{6} \\ \frac{3^{1/2}}{6} \\ \frac{2}{6} \\ \frac{5^{1/2}}{6} \\ \frac{6^{1/2}}{6} \\ \frac{5^{1/2}}{6} \\ \frac{2}{6} \\ \frac{3^{1/2}}{6} \\ \frac{2^{1/2}}{6} \\ \frac{1}{6} \end{vmatrix}$$

Now use the Greek letter notation to represent the state vector. So that's it; your state vector is

$$|\psi\rangle = \begin{vmatrix} \frac{1}{6} \\ \frac{2^{1/2}}{6} \\ \frac{3^{1/2}}{6} \\ \frac{2}{6} \\ \frac{5^{1/2}}{6} \\ \frac{6^{1/2}}{6} \\ \frac{5^{1/2}}{6} \\ \frac{2}{6} \\ \frac{3^{1/2}}{6} \\ \frac{2^{1/2}}{6} \\ \frac{1}{6} \end{vmatrix}$$

1. Assume you have two four-sided dice (in the shape of *tetrahedrons* — that is, mini pyramids). What are the relative probabilities of each state of the two dice? (***Note:*** Four-sided dice are odd to work with — the value of each die is represented by the number on the *bottom* face, because the dice can't come to rest on the top of a pyramid!)

Solve It

2. Put the relative probabilities of the various states of the four-sided dice into vector form.

Solve It

3. Convert the vector of relative probabilities in question 2 to probability amplitudes.

Solve It

4. Convert the relative probability amplitude vector you found for the four-sided dice in question 3 to a normalized state vector.

Solve It

Becoming a Notation Meister with Bras and Kets

Instead of writing out an entire vector each time, quantum physics usually uses a notation developed by physicist Paul Dirac — the *Dirac* or *bra-ket* notation. The two terms spell bra-ket, as in *bracket,* because when an operator appears between them, they bracket, or sandwich, that operator. Here's how write the two forms of state vectors:

- **Bras:** $\langle\psi|$
- **Kets:** $|\psi\rangle$

When you multiply the same state vector expressed as a bra and a ket together — the product is represented as $\langle\psi|\psi\rangle$ — you get 1. In other words, $\langle\psi|\psi\rangle = 1$. You get 1 because the sum of all the probabilities of being in the allowed states must equal 1.

If you have a bra, the corresponding ket is the *Hermitian conjugate* (which you get by taking the transpose and changing the sign of any imaginary values) of that bra — $\langle\psi|$ equals $|\psi\rangle^\dagger$ (where the † means the Hermitian conjugate). What does that mean in vector terms? Check out the following example.

Q. What's the bra for the state vector of a pair of dice? Verify that $\langle\psi|\psi\rangle = 1$.

A. $\left| \frac{1}{6} \quad \frac{2^{1/2}}{6} \quad \frac{3^{1/2}}{6} \quad \frac{2}{6} \quad \frac{5^{1/2}}{6} \quad \frac{6^{1/2}}{6} \quad \frac{5^{1/2}}{6} \quad \frac{2}{6} \quad \frac{3^{1/2}}{6} \quad \frac{2^{1/2}}{6} \quad \frac{1}{6} \right|$

Start with the ket:

$$|\psi\rangle = \begin{vmatrix} \frac{1}{6} \\ \frac{2^{1/2}}{6} \\ \frac{3^{1/2}}{6} \\ \frac{2}{6} \\ \frac{5^{1/2}}{6} \\ \frac{6^{1/2}}{6} \\ \frac{5^{1/2}}{6} \\ \frac{2}{6} \\ \frac{3^{1/2}}{6} \\ \frac{2^{1/2}}{6} \\ \frac{1}{6} \end{vmatrix}$$

Now find the complex conjugate of the ket. To do so in matrix terms, you take the transpose of the ket and then take the complex conjugate of each term (which does nothing in this case because all terms are real numbers). Finding the transpose just involves writing the columns of the ket as the rows of the bra, which gives you the following for the bra:

$$\begin{vmatrix} \frac{1}{6} & \frac{2^{1/2}}{6} & \frac{3^{1/2}}{6} & \frac{2}{6} & \frac{5^{1/2}}{6} & \frac{6^{1/2}}{6} & \frac{5^{1/2}}{6} & \frac{2}{6} & \frac{3^{1/2}}{6} & \frac{2^{1/2}}{6} & \frac{1}{6} \end{vmatrix}$$

To verify that $\langle\psi|\psi\rangle = 1$, multiply the bra and ket together using matrix multiplication like this:

$$\begin{vmatrix} \frac{1}{6} & \frac{2^{1/2}}{6} & \frac{3^{1/2}}{6} & \frac{2}{6} & \frac{5^{1/2}}{6} & \frac{6^{1/2}}{6} & \frac{5^{1/2}}{6} & \frac{2}{6} & \frac{3^{1/2}}{6} & \frac{2^{1/2}}{6} & \frac{1}{6} \end{vmatrix} \begin{vmatrix} \frac{1}{6} \\ \frac{2^{1/2}}{6} \\ \frac{3^{1/2}}{6} \\ \frac{2}{6} \\ \frac{5^{1/2}}{6} \\ \frac{6^{1/2}}{6} \\ \frac{5^{1/2}}{6} \\ \frac{2}{6} \\ \frac{3^{1/2}}{6} \\ \frac{2^{1/2}}{6} \\ \frac{1}{6} \end{vmatrix}$$

Complete the matrix multiplication to give you

$$\langle\psi|\psi\rangle = \frac{1}{36} + \frac{2}{36} + \frac{3}{36} + \frac{4}{36} + \frac{5}{36} + \frac{6}{36} + \frac{5}{36} + \frac{4}{36} + \frac{3}{36} + \frac{2}{36} + \frac{1}{36} = 1$$

5. Find the bra for the state vector of a pair of four-sided dice.

Solve It

6. Confirm that $\langle\psi|\psi\rangle$ for the bra and ket for the four-sided dice equals 1.

Solve It

Getting into the Big Leagues with Operators

What are bras and kets useful for? They represent a system in a stateless way — that is, you don't have to know which state every element in a general ket or bra corresponds to; you don't have to spell out each vector. Therefore, you can use kets and bras in a general way to work with systems. In other words, you can do a lot of math on kets and bras that would be unwieldy if you had to spell out all the elements of a state vector every time. Operators can assist you. This section takes a closer look at how you can use operators to make your calculations.

Introducing operators and getting into a healthy, orthonormal relationship

Kets and bras describe the state of a system. But what if you want to measure some quantity of the system (such as its momentum) or change the system (such as by raiding a hydrogen atom to an excited state)? That's where operators come in. You apply an *operator* to a bra or ket to extract a value and/or change the bra or ket to a different state. In general, an operator gives you a new bra or ket when you use that operator: $A|\psi\rangle = |\psi'\rangle$.

Some of the most important operators you need to know include the following:

- **Hamiltonian operator:** Designated as H, this operator is the most important in quantum physics. When applied to a bra or ket, it gives you the energy of the state that the bra or ket represents (as a constant) multiplied by that bra or ket again:

  $$H|\psi\rangle = E|\psi\rangle$$

  E is the energy of the particle represented by the ket $|\psi\rangle$.

- **Unity or identity operator:** Designated as I, this operator leaves kets unchanged:

  $$I|\psi\rangle = |\psi\rangle$$

- **Gradient operator:** Designated as ∇, this operator takes the derivative. It works like this:

  $$\nabla|\psi\rangle = \frac{\partial}{\partial x}|\psi\rangle \boldsymbol{i} + \frac{\partial}{\partial y}|\psi\rangle \boldsymbol{j} + \frac{\partial}{\partial z}|\psi\rangle \boldsymbol{k}$$

- **Linear momentum operator:** Designated as P, this operator finds the momentum of a state. It looks like this:

  $$P|\psi\rangle = -i\hbar\nabla|\psi\rangle$$

- **Laplacian operator:** Designated as Δ, or ∇^2, this operator is much like a second-order gradient, which means it takes the second derivative. It looks like this:

  $$\nabla^2|\psi\rangle = \Delta|\psi\rangle = \frac{\partial^2}{\partial x^2}|\psi\rangle + \frac{\partial^2}{\partial y^2}|\psi\rangle + \frac{\partial^2}{2z^2}|\psi\rangle$$

In general, multiplying operators together is not the same independent of order, so for the operators A and B,

$$AB \neq BA$$

You can find the complex conjugate of an operator A, denoted $A^\dagger$, like this:

$$\langle\psi|A^\dagger|\phi\rangle = \langle\phi|A|\psi\rangle$$

When working with kets and bras, keep the following in mind:

- Two kets, $|\psi\rangle$ and $|\phi\rangle$, are said to be *orthogonal* if

  $$\langle\psi|\phi\rangle$$

- Two kets are said to be *orthonormal* if all three of the following apply:
  - $\langle\psi|\phi\rangle = 0$
  - $\langle\psi|\psi\rangle = 1$
  - $\langle\phi|\phi\rangle = 1$

Q. Find an orthonormal ket to the bra $\left|\frac{1}{2^{1/2}} \quad 0 \quad 0 \quad \frac{1}{2^{1/2}}\right|$.

A. $\begin{vmatrix} \frac{1}{2^{1/2}} \\ 0 \\ 0 \\ -\frac{-1}{2^{1/2}} \end{vmatrix}$

You know that to be orthonormal, the following relations must be true:

- $\langle\psi|\phi\rangle = 0$
- $\langle\psi|\psi\rangle = 1$

So you need to construct a ket made up of elements A, B, C, D such that

$$\left|\frac{1}{2^{1/2}} \quad 0 \quad 0 \quad \frac{1}{2^{1/2}}\right| \begin{vmatrix} A \\ B \\ C \\ D \end{vmatrix} = 0$$

Do the matrix multiplication to get

$$\frac{A}{2^{1/2}} + \frac{D}{2^{1/2}} = 0$$

So therefore, A = –D (and you can leave B and C at 0; their value is arbitrary because you multiply them by the zeroes in the bra, giving you a product of 0).

You're not free to choose just any values for A and D because $\langle\psi|\psi\rangle$ must equal 1. So you can choose $A = \frac{1}{2^{1/2}}$ and $D = -\frac{1}{2^{1/2}}$, giving you the following ket:

$$\begin{vmatrix} \frac{1}{2^{1/2}} \\ 0 \\ 0 \\ -\frac{1}{2^{1/2}} \end{vmatrix}$$

7. Find an orthonormal ket to the bra

$$\left| \frac{1}{2^{1/2}} \quad \frac{1}{2^{1/2}} \right|$$

Solve It

8. Find the identity operator for bras and kets with six elements.

Solve It

Grasping Hermitian operators and adjoints

Operators that are equal to their Hermitian adjoints are called *Hermitian operators*. In other words, an operator is Hermitian if

$$A^{\dagger} = A$$

Here's how you find the Hermitian adjoint of an operator, A:

1. **Find the transpose by interchanging the rows and columns, A^T.**
2. **Take the complex conjugate.**

$$A^{T^*} = A^{\dagger}$$

In addition, finding the inverse is often useful because applying the inverse of an operator undoes the work the operator did: $A^{-1}A = AA^{-1} = I$. For instance, when you have equations like $Ax = y$, solving for x is easy if you can find the inverse of A: $x = A^{-1}y$. But finding the inverse of a large matrix usually isn't easy, so quantum physics calculations are sometimes limited to working with *unitary operators*, U, where the operator's inverse is equal to its Hermitian adjoint:

$$U^{-1} = U^{\dagger}$$

Getting Physical Measurements with Expectation Values

Everything in quantum physics is done in terms of probabilities, so making predictions becomes very important. The biggest such prediction is the expectation value. The *expectation value* of an operator is the average value the operator will give you when you apply it to a particular system many times.

The expectation value is a weighted mean of the probable values of an operator. Here's how you'd find the expectation value of an operator A:

$$\text{Expectation value} = \langle \psi | A | \psi \rangle$$

Because you can express $\langle \psi |$ as a row vector and $| \psi \rangle$ as a column vector, you can express the operator A as a square matrix.

Finding the expectation value is so common that you often find $\langle \psi | A | \psi \rangle$ abbreviated as $\langle A \rangle$.

The expression $|\phi\rangle\langle\psi|$ is actually a linear operator. To see that, apply $|\phi\rangle\langle\psi|$ to a ket, $|\chi\rangle$:

$$|\phi\rangle\langle\psi|\chi\rangle$$

which is $\langle\psi|\chi\rangle|\phi\rangle$.

The expression $\langle\psi|\chi\rangle$ is always a complex number (which could be purely real), so this breaks down to $c|\phi\rangle$, where c is a complex number, so $|\phi\rangle\langle\psi|$ is indeed a linear operator.

Q. What is the expectation value of rolling two dice?

A. **Seven.** For two dice, the expectation value is a sum of terms, and each term is a value that the dice can display multiplied by the probability that that value will appear. The bra and ket handle the probabilities, so the operator you create for this problem, which I call the A operator for this example, needs to store the dice values (2 through 12) for each probability. Therefore, the operator A looks like this:

$$A = \begin{vmatrix} 2 & 0 & 0 & 0 & 0 & 0 & 0 & 0 & 0 & 0 & 0 \\ 0 & 3 & 0 & 0 & 0 & 0 & 0 & 0 & 0 & 0 & 0 \\ 0 & 0 & 4 & 0 & 0 & 0 & 0 & 0 & 0 & 0 & 0 \\ 0 & 0 & 0 & 5 & 0 & 0 & 0 & 0 & 0 & 0 & 0 \\ 0 & 0 & 0 & 0 & 6 & 0 & 0 & 0 & 0 & 0 & 0 \\ 0 & 0 & 0 & 0 & 0 & 7 & 0 & 0 & 0 & 0 & 0 \\ 0 & 0 & 0 & 0 & 0 & 0 & 8 & 0 & 0 & 0 & 0 \\ 0 & 0 & 0 & 0 & 0 & 0 & 0 & 9 & 0 & 0 & 0 \\ 0 & 0 & 0 & 0 & 0 & 0 & 0 & 0 & 10 & 0 & 0 \\ 0 & 0 & 0 & 0 & 0 & 0 & 0 & 0 & 0 & 11 & 0 \\ 0 & 0 & 0 & 0 & 0 & 0 & 0 & 0 & 0 & 0 & 12 \end{vmatrix}$$

To find the expectation value of A, you need to calculate $\langle\psi|A|\psi\rangle$. Spelling that out in terms of components gives you the following:

$$\langle\psi|A|\psi\rangle = \begin{vmatrix} \frac{1}{6} & \frac{2^{1/2}}{6} & \frac{3^{1/2}}{6} & \frac{2}{6} & \frac{5^{1/2}}{6} & \frac{6^{1/2}}{6} & \frac{5^{1/2}}{6} & \frac{2}{6} & \frac{3^{1/2}}{6} & \frac{2^{1/2}}{6} & \frac{1}{6} \end{vmatrix} \begin{vmatrix} 2 & 0 & 0 & 0 & 0 & 0 & 0 & 0 & 0 & 0 & 0 \\ 0 & 3 & 0 & 0 & 0 & 0 & 0 & 0 & 0 & 0 & 0 \\ 0 & 0 & 4 & 0 & 0 & 0 & 0 & 0 & 0 & 0 & 0 \\ 0 & 0 & 0 & 5 & 0 & 0 & 0 & 0 & 0 & 0 & 0 \\ 0 & 0 & 0 & 0 & 6 & 0 & 0 & 0 & 0 & 0 & 0 \\ 0 & 0 & 0 & 0 & 0 & 7 & 0 & 0 & 0 & 0 & 0 \\ 0 & 0 & 0 & 0 & 0 & 0 & 8 & 0 & 0 & 0 & 0 \\ 0 & 0 & 0 & 0 & 0 & 0 & 0 & 9 & 0 & 0 & 0 \\ 0 & 0 & 0 & 0 & 0 & 0 & 0 & 0 & 10 & 0 & 0 \\ 0 & 0 & 0 & 0 & 0 & 0 & 0 & 0 & 0 & 11 & 0 \\ 0 & 0 & 0 & 0 & 0 & 0 & 0 & 0 & 0 & 0 & 12 \end{vmatrix} \begin{vmatrix} \frac{1}{6} \\ \frac{2^{1/2}}{6} \\ \frac{3^{1/2}}{6} \\ \frac{2}{6} \\ \frac{5^{1/2}}{6} \\ \frac{6^{1/2}}{6} \\ \frac{5^{1/2}}{6} \\ \frac{2}{6} \\ \frac{3^{1/2}}{6} \\ \frac{2^{1/2}}{6} \\ \frac{1}{6} \end{vmatrix}$$

Doing the matrix multiplication gives you

$$\langle \psi | A | \psi \rangle = 7$$

So the expectation value of a roll of the dice is 7.

9. Find the expectation value of two four-sided dice.

Solve It

10. Find the expectation value of the identity operator for a pair of normal, six-sided dice (see the earlier section "Introducing operators and getting into a healthy, orthonormal relationship" for more on the identity operator).

Solve It

Commutators: Checking How Different Operators Really Are

In quantum physics, the measure of the difference between applying operator A and then B, versus B and then A, is called the operators' *commutator.* If two operators have a commutator that's 0, they *commute,* and the order in which you apply them doesn't make any difference. In other words, operators that commute don't interfere with each other, and that's useful to know when you're working with multiple operators. You can independently use commuting operators, whereas you can't independently use noncommuting ones.

Here's how you define the commutator of operators A and B:

$$[A, B] = AB - BA$$

Two operators commute with each other if their commutator is equal to 0:

$$[A, B] = 0$$

The Hermitian adjoint of a commutator works this way:

$$[A, B]^\dagger = [B^\dagger, A^\dagger]$$

Check out the following example, which illustrates the concept of commuting.

Q. Show that any operator commutes with itself.

A. **[A, A] = 0.** The definition of a commutator is [A, B] = AB – BA. And if both operators are A, you get

$$[A, A] = AA - AA$$

But AA – AA = 0, so you get

$$[A, A] = AA - AA = 0$$

11. What is [A, B] in terms of [B, A]?

Solve It

12. What is the Hermitian adjoint of a commutator $[A, B]^\dagger$ if A and B are Hermitian operators?

Solve It

Simplifying Matters by Finding Eigenvectors and Eigenvalues

When you apply an operator to a ket, you generally get a new ket. For instance, $A|\psi\rangle = |\psi'\rangle$. However, sometimes you can make matters a little simpler by casting your problem in terms of eigenvectors and eigenvalues (*eigen* is German for "innate" or "natural"). Instead of giving you an entirely new ket, applying an operator to its *eigenvector* (a ket) merely gives you the same eigenvector back again, multiplied by its *eigenvalue* (a constant). In other words, $|\psi\rangle$ is an eigenvector of the operator A if the number a is a complex constant and $A|\psi\rangle = a|\psi\rangle$.

So applying A to one of its eigenvectors, $|\psi\rangle$, gives you $|\psi\rangle$ back, multiplied by that eigenvector's eigenvalue, a. An eigenvalue can be complex, but note that if the operators are Hermitian, the values of a are real and their eigenvectors are orthogonal (see the earlier section "Grasping Hermitian operators and adjoints" for more on Hermitian operators).

To find an operator's eigenvalues, you want to find a, such that

$$A|\psi\rangle = a|\psi\rangle$$

You can rewrite the equation this way, where I is the identity matrix (that is, it contains all 0s except for the 1s running along the diagonal from upper left to lower right):

$$(A - aI)|\psi\rangle = 0$$

For this equation to have a solution, the matrix determinant of $(A - aI)$ must equal 0:

$$\det(A - aI) = 0$$

Solving this relation gives you an equation for a — and the roots of the equation are the eigenvalues. You then plug the eigenvalues, one by one, into the equation $(A - aI)|\psi\rangle = 0$ to find the eigenvectors.

If two or more of the eigenvalues are the same, that eigenvalue is said to be *degenerate*.

Know that many systems, like free particles, don't have a number of set discrete energy states; their states are continuous. In such circumstances, you move from a state vector like $|\psi\rangle$ to a continuous wave function, $\psi(\boldsymbol{r})$. How does $\psi(\boldsymbol{r})$ relate to $|\psi\rangle$? You have to relate the stateless vector $|\psi\rangle$ to normal spatial dimensions, which you do with a state vector where the states correspond to possible positions, $\langle\boldsymbol{r}|$ (see my book *Quantum Physics For Dummies* [Wiley] for all the details). In that case, $\psi(\boldsymbol{r}) = \langle\boldsymbol{r}|\psi\rangle$.

Q. What are the eigenvectors and eigenvalues of the following operator, which presents the operator for two six-sided dice?

$$\begin{vmatrix} 2 & 0 & 0 & 0 & 0 & 0 & 0 & 0 & 0 & 0 & 0 \\ 0 & 3 & 0 & 0 & 0 & 0 & 0 & 0 & 0 & 0 & 0 \\ 0 & 0 & 4 & 0 & 0 & 0 & 0 & 0 & 0 & 0 & 0 \\ 0 & 0 & 0 & 5 & 0 & 0 & 0 & 0 & 0 & 0 & 0 \\ 0 & 0 & 0 & 0 & 6 & 0 & 0 & 0 & 0 & 0 & 0 \\ 0 & 0 & 0 & 0 & 0 & 7 & 0 & 0 & 0 & 0 & 0 \\ 0 & 0 & 0 & 0 & 0 & 0 & 8 & 0 & 0 & 0 & 0 \\ 0 & 0 & 0 & 0 & 0 & 0 & 0 & 9 & 0 & 0 & 0 \\ 0 & 0 & 0 & 0 & 0 & 0 & 0 & 0 & 10 & 0 & 0 \\ 0 & 0 & 0 & 0 & 0 & 0 & 0 & 0 & 0 & 11 & 0 \\ 0 & 0 & 0 & 0 & 0 & 0 & 0 & 0 & 0 & 0 & 12 \end{vmatrix}$$

A. **The eigenvalues are 2, 3, 4, 5, ..., 12, and the eigenvectors are**

$$\begin{vmatrix} 1 \\ 0 \\ 0 \\ 0 \\ 0 \\ 0 \\ 0 \\ 0 \\ 0 \\ 0 \\ 0 \end{vmatrix} \begin{vmatrix} 0 \\ 1 \\ 0 \\ 0 \\ 0 \\ 0 \\ 0 \\ 0 \\ 0 \\ 0 \\ 0 \end{vmatrix} \ldots \begin{vmatrix} 0 \\ 0 \\ 0 \\ 0 \\ 0 \\ 0 \\ 0 \\ 0 \\ 0 \\ 0 \\ 1 \end{vmatrix}$$

Here's the operator you want to find the eigenvalues and eigenvectors of:

$$\begin{vmatrix} 2 & 0 & 0 & 0 & 0 & 0 & 0 & 0 & 0 & 0 & 0 \\ 0 & 3 & 0 & 0 & 0 & 0 & 0 & 0 & 0 & 0 & 0 \\ 0 & 0 & 4 & 0 & 0 & 0 & 0 & 0 & 0 & 0 & 0 \\ 0 & 0 & 0 & 5 & 0 & 0 & 0 & 0 & 0 & 0 & 0 \\ 0 & 0 & 0 & 0 & 6 & 0 & 0 & 0 & 0 & 0 & 0 \\ 0 & 0 & 0 & 0 & 0 & 7 & 0 & 0 & 0 & 0 & 0 \\ 0 & 0 & 0 & 0 & 0 & 0 & 8 & 0 & 0 & 0 & 0 \\ 0 & 0 & 0 & 0 & 0 & 0 & 0 & 9 & 0 & 0 & 0 \\ 0 & 0 & 0 & 0 & 0 & 0 & 0 & 0 & 10 & 0 & 0 \\ 0 & 0 & 0 & 0 & 0 & 0 & 0 & 0 & 0 & 11 & 0 \\ 0 & 0 & 0 & 0 & 0 & 0 & 0 & 0 & 0 & 0 & 12 \end{vmatrix}$$

This operator operates in 11-dimensional space, so you need to find 11 eigenvectors and 11 corresponding eigenvalues.

This operator is already diagonal, so this problem is easy — just take unit vectors in the 11 different directions of the eigenvectors. Here's what the first eigenvector is:

$$\begin{vmatrix} 1 \\ 0 \\ 0 \\ 0 \\ 0 \\ 0 \\ 0 \\ 0 \\ 0 \\ 0 \\ 0 \end{vmatrix}$$

And here's the second eigenvector:

$$\begin{vmatrix} 0 \\ 1 \\ 0 \\ 0 \\ 0 \\ 0 \\ 0 \\ 0 \\ 0 \\ 0 \\ 0 \end{vmatrix}$$

And so on, up to the 11th eigenvector

$$\begin{vmatrix} 0 \\ 0 \\ 0 \\ 0 \\ 0 \\ 0 \\ 0 \\ 0 \\ 0 \\ 0 \\ 1 \end{vmatrix}$$

What about the eigenvalues? The eigenvalues are the values you get when you apply the operator to an eigenvector, and because the eigenvectors are just unit vectors in all 11 dimensions, the eigenvalues are the numbers on the diagonal of the operator — that is, 2, 3, 4, and so on, up to 12.

13. What are the eigenvalues and eigenvectors of this operator?

$$A = \begin{vmatrix} 2 & 1 \\ 1 & 2 \end{vmatrix}$$

Solve It

14. What are the eigenvalues and eigenvectors of this operator?

$$A = \begin{vmatrix} 3 & -1 \\ 4 & -2 \end{vmatrix}$$

Solve It

Answers to Problems on State Vectors

The following are the answers to the practice questions presented earlier in this chapter. I first repeat the problems and give the answers in bold. Then you can see the answers worked out, step by step.

1. Assume you have two four-sided dice (in the shape of *tetrahedons* — that is, mini pyramids). What are the relative probabilities of each state of the two dice? **Here's the answer:**

   1 = Relative probability of getting a 2

   2 = Relative probability of getting a 3

   3 = Relative probability of getting a 4

   4 = Relative probability of getting a 5

   3 = Relative probability of getting a 6

   2 = Relative probability of getting a 7

   1 = Relative probability of getting a 8

   Adding up the various totals of the two four-sided dice gives you the number of ways each total can appear, and that's the relative probability of each state.

2. Put the relative probabilities of the various states of the four-sided dice into vector form.

$$\begin{vmatrix} 1 \\ 2 \\ 3 \\ 4 \\ 3 \\ 2 \\ 1 \end{vmatrix}$$

   Just assemble the relative probabilities of each state into vector format.

3. Convert the vector of relative probabilities in question 2 to probability amplitudes.

$$\begin{vmatrix} 1 \\ 2^{1/2} \\ 3^{1/2} \\ 2 \\ 3^{1/2} \\ 2^{1/2} \\ 1 \end{vmatrix}$$

   To find the probability amplitudes, just take the square root of the relative probabilities.

4 Convert the relative probability amplitude vector you found for the four-sided dice in question 3 to a normalized state vector.

$$\begin{vmatrix} \frac{1}{4} \\ \frac{2^{1/2}}{4} \\ \frac{3^{1/2}}{4} \\ \frac{1}{2} \\ \frac{3^{1/2}}{4} \\ \frac{2^{1/2}}{4} \\ \frac{1}{4} \end{vmatrix}$$

To normalize the state vector, divide each term by the square root of the sum of the squares of each term: $1^2 + (2^{1/2})^2 + (3^{1/2})^2 + 2^2 + (3^{1/2})^2 + (2^{1/2})^2 + 1^2 = 1 + 2 + 3 + 4 + 3 + 2 + 1 = 16$, and $16^{1/2} = 4$, so divide each term by 4. Doing so ensures that the square of the state vector gives you a total value of 1.

5 Find the bra for the state vector of a pair of four-sided dice. **The answer is**

$$\begin{vmatrix} \frac{1}{4} & \frac{2^{1/2}}{4} & \frac{3^{1/2}}{4} & \frac{1}{2} & \frac{3^{1/2}}{4} & \frac{2^{1/2}}{4} & \frac{1}{4} \end{vmatrix}$$

To find the bra, start with the ket that you already found in problem 4:

$$\begin{vmatrix} \frac{1}{4} \\ \frac{2^{1/2}}{4} \\ \frac{3^{1/2}}{4} \\ \frac{1}{2} \\ \frac{3^{1/2}}{4} \\ \frac{2^{1/2}}{4} \\ \frac{1}{4} \end{vmatrix}$$

Take the transpose of the ket and the complex conjugate of each term (which does nothing, because each term is real) to get

$$\begin{vmatrix} \frac{1}{4} & \frac{2^{1/2}}{4} & \frac{3^{1/2}}{4} & \frac{1}{2} & \frac{3^{1/2}}{4} & \frac{2^{1/2}}{4} & \frac{1}{4} \end{vmatrix}$$

6 Confirm that $\langle\psi|\psi\rangle$ for the bra and ket for the four-sided dice equals 1. **Here's the answer:** $\boldsymbol{\langle\psi|\psi\rangle = 1}$

To find $\langle\psi|\psi\rangle$, perform this multiplication:

$$\begin{vmatrix} \frac{1}{4} & \frac{2^{1/2}}{4} & \frac{3^{1/2}}{4} & \frac{1}{2} & \frac{3^{1/2}}{4} & \frac{2^{1/2}}{4} & \frac{1}{4} \end{vmatrix} \begin{vmatrix} \frac{1}{4} \\ \frac{2^{1/2}}{4} \\ \frac{3^{1/2}}{4} \\ \frac{1}{2} \\ \frac{3^{1/2}}{4} \\ \frac{2^{1/2}}{4} \\ \frac{1}{4} \end{vmatrix}$$

To find $\langle\psi|\psi\rangle$, perform this multiplication, giving you 1:

$$\langle\psi|\psi\rangle = \frac{1}{16} + \frac{2}{16} + \frac{3}{16} + \frac{4}{16} + \frac{3}{16} + \frac{2}{16} + \frac{1}{16} = 1$$

7 Find an orthonormal ket to the bra $\begin{vmatrix} \frac{1}{2^{1/2}} & \frac{1}{2^{1/2}} \end{vmatrix}$.

$$\begin{vmatrix} \boldsymbol{\frac{1}{2^{1/2}}} \\ \boldsymbol{-\frac{1}{2^{1/2}}} \end{vmatrix}$$

You know that to be orthonormal, the following relations must be true:

- $\langle\psi|\phi\rangle = 0$
- $\langle\psi|\psi\rangle = 1$

So you need to construct a ket made up of elements A and B such that

$$\begin{vmatrix} \frac{1}{2^{1/2}} & \frac{1}{2^{1/2}} \end{vmatrix} \begin{vmatrix} A \\ B \end{vmatrix} = 0$$

Do the matrix multiplication to get

$$\frac{A}{2^{1/2}} + \frac{D}{2^{1/2}} = 0$$

So therefore, A = –D. You're not free to choose just any values for A and D because $\langle\psi|\psi\rangle$ must equal 1. So you can choose $A = \frac{1}{2^{1/2}}$ and $D = \frac{-1}{2^{1/2}}$ to make the math come out right here, giving you the following ket:

$$\begin{vmatrix} \frac{1}{2^{1/2}} \\ \frac{-1}{2^{1/2}} \end{vmatrix}$$

8 Find the identity operator for bras and kets with six elements.

$$\begin{vmatrix} \mathbf{1} & \mathbf{0} & \mathbf{0} & \mathbf{0} & \mathbf{0} & \mathbf{0} \\ \mathbf{0} & \mathbf{1} & \mathbf{0} & \mathbf{0} & \mathbf{0} & \mathbf{0} \\ \mathbf{0} & \mathbf{0} & \mathbf{1} & \mathbf{0} & \mathbf{0} & \mathbf{0} \\ \mathbf{0} & \mathbf{0} & \mathbf{0} & \mathbf{1} & \mathbf{0} & \mathbf{0} \\ \mathbf{0} & \mathbf{0} & \mathbf{0} & \mathbf{0} & \mathbf{1} & \mathbf{0} \\ \mathbf{0} & \mathbf{0} & \mathbf{0} & \mathbf{0} & \mathbf{0} & \mathbf{1} \end{vmatrix}$$

You need a matrix I such that

$$I|\psi\rangle = |\psi\rangle$$

In this problem, you're working with bras and kets with six elements:

$$\begin{vmatrix} A \\ B \\ C \\ D \\ E \\ F \end{vmatrix}$$

Therefore, you need a matrix that looks like this:

$$\begin{vmatrix} 1 & 0 & 0 & 0 & 0 & 0 \\ 0 & 1 & 0 & 0 & 0 & 0 \\ 0 & 0 & 1 & 0 & 0 & 0 \\ 0 & 0 & 0 & 1 & 0 & 0 \\ 0 & 0 & 0 & 0 & 1 & 0 \\ 0 & 0 & 0 & 0 & 0 & 1 \end{vmatrix}$$

9 Find the expectation value of two four-sided dice. **The answer is 5.**

For two four-sided dice, the expectation value is a sum of terms, and each term is a value that the dice can display, multiplied by the probability that that value will appear.

The bra and ket will handle the probabilities, so it's up to the operator you create for this — call it the A operator — to store the dice values (2 through 8) for each probability, which means that the operator A looks like this:

$$A = \begin{vmatrix} 2 & 0 & 0 & 0 & 0 & 0 & 0 \\ 0 & 3 & 0 & 0 & 0 & 0 & 0 \\ 0 & 0 & 4 & 0 & 0 & 0 & 0 \\ 0 & 0 & 0 & 5 & 0 & 0 & 0 \\ 0 & 0 & 0 & 0 & 6 & 0 & 0 \\ 0 & 0 & 0 & 0 & 0 & 7 & 0 \\ 0 & 0 & 0 & 0 & 0 & 0 & 8 \end{vmatrix}$$

To find the expectation value of A, you need to calculate $\langle\psi|A|\psi\rangle$. Spelling that out in terms of components gives you the following:

$$\langle\psi|A|\psi\rangle = \begin{vmatrix} \frac{1}{4} & \frac{2^{1/2}}{4} & \frac{3^{1/2}}{4} & \frac{1}{2} & \frac{3^{1/2}}{4} & \frac{2^{1/2}}{4} & \frac{1}{4} \end{vmatrix} \begin{vmatrix} 2 & 0 & 0 & 0 & 0 & 0 & 0 \\ 0 & 3 & 0 & 0 & 0 & 0 & 0 \\ 0 & 0 & 4 & 0 & 0 & 0 & 0 \\ 0 & 0 & 0 & 5 & 0 & 0 & 0 \\ 0 & 0 & 0 & 0 & 6 & 0 & 0 \\ 0 & 0 & 0 & 0 & 0 & 7 & 0 \\ 0 & 0 & 0 & 0 & 0 & 0 & 8 \end{vmatrix} \begin{vmatrix} \frac{1}{4} \\ \frac{2^{1/2}}{4} \\ \frac{3^{1/2}}{4} \\ \frac{1}{2} \\ \frac{3^{1/2}}{4} \\ \frac{2^{1/2}}{4} \\ \frac{1}{4} \end{vmatrix}$$

Doing the matrix multiplication gives you

$$\langle\psi|A|\psi\rangle = 5$$

So the expectation value of a roll of the pair of four-sided dice is 5.

10 Find the expectation value of the identity operator for a pair of normal, six-sided dice. **The answer is 1.**

For two dice, the expectation value is a sum of terms, and each term is a value that the dice can display, multiplied by the probability that that value will appear. The bra and ket will handle the probabilities. The operator A is the identity operator, so it looks like this:

$$A = \begin{vmatrix} 1 & 0 & 0 & 0 & 0 & 0 & 0 & 0 & 0 & 0 & 0 \\ 0 & 1 & 0 & 0 & 0 & 0 & 0 & 0 & 0 & 0 & 0 \\ 0 & 0 & 1 & 0 & 0 & 0 & 0 & 0 & 0 & 0 & 0 \\ 0 & 0 & 0 & 1 & 0 & 0 & 0 & 0 & 0 & 0 & 0 \\ 0 & 0 & 0 & 0 & 1 & 0 & 0 & 0 & 0 & 0 & 0 \\ 0 & 0 & 0 & 0 & 0 & 1 & 0 & 0 & 0 & 0 & 0 \\ 0 & 0 & 0 & 0 & 0 & 0 & 1 & 0 & 0 & 0 & 0 \\ 0 & 0 & 0 & 0 & 0 & 0 & 0 & 1 & 0 & 0 & 0 \\ 0 & 0 & 0 & 0 & 0 & 0 & 0 & 0 & 1 & 0 & 0 \\ 0 & 0 & 0 & 0 & 0 & 0 & 0 & 0 & 0 & 1 & 0 \\ 0 & 0 & 0 & 0 & 0 & 0 & 0 & 0 & 0 & 0 & 1 \end{vmatrix}$$

To find the expectation value of I, you need to calculate $\langle \psi | A | \psi \rangle$. Spelling that out in terms of components gives you the following:

$$\langle \psi | I | \psi \rangle =$$

$$\begin{vmatrix} \frac{1}{6} & \frac{2^{1/2}}{6} & \frac{3^{1/2}}{6} & \frac{2}{6} & \frac{5^{1/2}}{6} & \frac{6^{1/2}}{6} & \frac{5^{1/2}}{6} & \frac{2}{6} & \frac{3^{1/2}}{6} & \frac{2^{1/2}}{6} & \frac{1}{6} \end{vmatrix} \begin{vmatrix} 1 & 0 & 0 & 0 & 0 & 0 & 0 & 0 & 0 & 0 & 0 \\ 0 & 1 & 0 & 0 & 0 & 0 & 0 & 0 & 0 & 0 & 0 \\ 0 & 0 & 1 & 0 & 0 & 0 & 0 & 0 & 0 & 0 & 0 \\ 0 & 0 & 0 & 1 & 0 & 0 & 0 & 0 & 0 & 0 & 0 \\ 0 & 0 & 0 & 0 & 1 & 0 & 0 & 0 & 0 & 0 & 0 \\ 0 & 0 & 0 & 0 & 0 & 1 & 0 & 0 & 0 & 0 & 0 \\ 0 & 0 & 0 & 0 & 0 & 0 & 1 & 0 & 0 & 0 & 0 \\ 0 & 0 & 0 & 0 & 0 & 0 & 0 & 1 & 0 & 0 & 0 \\ 0 & 0 & 0 & 0 & 0 & 0 & 0 & 0 & 1 & 0 & 0 \\ 0 & 0 & 0 & 0 & 0 & 0 & 0 & 0 & 0 & 1 & 0 \\ 0 & 0 & 0 & 0 & 0 & 0 & 0 & 0 & 0 & 0 & 1 \end{vmatrix} \begin{vmatrix} \frac{1}{6} \\ \frac{2^{1/2}}{6} \\ \frac{3^{1/2}}{6} \\ \frac{2}{6} \\ \frac{5^{1/2}}{6} \\ \frac{6^{1/2}}{6} \\ \frac{5^{1/2}}{6} \\ \frac{2}{6} \\ \frac{3^{1/2}}{6} \\ \frac{2^{1/2}}{6} \\ \frac{1}{6} \end{vmatrix}$$

Doing the matrix multiplication gives you

$$\langle\psi|\mathrm{I}|\psi\rangle = 1$$

So the expectation value of the identity operator is 1.

11 What is [A, B] in terms of [B, A]? **The answer is [A, B] = –[B, A].**

The definition of a commutator [A, B] is [A, B] = AB – BA. And [B, A] = BA – AB, or [B, A] = –AB + BA. You can write this as

$$[B, A] = -(AB - BA)$$

But AB – BA = [A, B], so [B, A] = –[A, B], or

$$[A, B] = -[B, A]$$

12 What is the Hermitian adjoint of a commutator $[A, B]^\dagger$ if A and B are Hermitian operators? **The answer is $[A, B]^\dagger = -[A, B]$.**

You want to figure out what the following expression is:

$$[A, B]^\dagger$$

Expanding gives you

$$[A, B]^\dagger = (AB - BA)^\dagger$$

And expanding this gives you

$$[A, B]^\dagger = (AB - BA)^\dagger = B^\dagger A^\dagger - A^\dagger B^\dagger$$

For Hermitian operators, $A = A^\dagger$, so

$$[A, B]^\dagger = B^\dagger A^\dagger - A^\dagger B^\dagger = BA - AB$$

And BA – AB is just –[A, B], so you get

$$[A, B]^\dagger = -[A, B]$$

where A and B are Hermitian operators.

Note that when you take the Hermitian adjoint of an expression and get the same thing back with a negative sign in front of it, the expression is called *anti-Hermitian,* so the commutator of two Hermitian operators is anti-Hermitian.

13 What are the eigenvalues and eigenvectors of this operator?

$$A = \begin{vmatrix} 2 & 1 \\ 1 & 2 \end{vmatrix}$$

The eigenvalues of A are a_1 = 1 and a_2 = 3. The eigenvectors are

$$\begin{vmatrix} 1 \\ -1 \end{vmatrix} \textbf{ and } \begin{vmatrix} 1 \\ 1 \end{vmatrix}$$

First find A – *a*I:

$$A - \lambda I = \begin{vmatrix} 2-a & 1 \\ 1 & 2-a \end{vmatrix}$$

Now find the determinant:

$$\det(A - aI) = (2-a)(2-a) - 1$$
$$= a^2 - 4a + 3$$

Factor this into

$$\det(A - aI) = a^2 - 4a + 3 = (a-1)(a-3)$$

So the eigenvalues of A are $a_1 = 1$ and $a_2 = 3$.

To find the eigenvector corresponding to a_1, substitute a_1 into A – *a*I:

$$A - aI = \begin{vmatrix} 1 & 1 \\ 1 & 1 \end{vmatrix}$$

Because (A – *a*I)*x* = 0, you have

$$\begin{vmatrix} 1 & 1 \\ 1 & 1 \end{vmatrix}\begin{vmatrix} x_1 \\ x_2 \end{vmatrix} = \begin{vmatrix} 0 \\ 0 \end{vmatrix}$$

Because every row of this matrix equation must be true, you know that $x_1 = -x_2$. And that means that up to an arbitrary constant, the eigenvector corresponding to a_1 is

$$c\begin{vmatrix} 1 \\ -1 \end{vmatrix}$$

Drop the arbitrary constant, and just write this as

$$\begin{vmatrix} 1 \\ -1 \end{vmatrix}$$

How about the eigenvector corresponding to a_2? Plugging a_2 in gives you

$$A - aI = \begin{vmatrix} -1 & 1 \\ 1 & -1 \end{vmatrix}$$

Then you have

$$\begin{vmatrix} -1 & 1 \\ 1 & -1 \end{vmatrix}\begin{vmatrix} x_1 \\ x_2 \end{vmatrix} = \begin{vmatrix} 0 \\ 0 \end{vmatrix}$$

So $x_1 = 0$, and that means that up to an arbitrary constant, the eigenvector corresponding to a_2 is

$$c\begin{vmatrix} 1 \\ 1 \end{vmatrix}$$

Drop the arbitrary constant and just write this as

$$\begin{vmatrix} 1 \\ 1 \end{vmatrix}$$

14 What are the eigenvalues and eigenvectors of this operator?

$$A = \begin{vmatrix} 3 & -1 \\ 4 & -2 \end{vmatrix}$$

The eigenvalues of A are $a_1 = 2$ and $a_2 = -1$. The eigenvectors are

$$\begin{vmatrix} 1 \\ 1 \end{vmatrix} \textbf{ and } \begin{vmatrix} 1 \\ 4 \end{vmatrix}$$

First, find A – aI:

$$A - aI = \begin{vmatrix} 3-a & -1 \\ 4 & -2-a \end{vmatrix}$$

Now find the determinant:

$$\det(A - aI) = (3-a)(-2-a) + 4$$
$$= a^2 - a - 2$$

Factor this into

$$\det(A - aI) = a^2 - a - 2 = (a+1)(a-2)$$

So the eigenvalues of A are $a_1 = 2$ and $a_2 = -1$.

To find the eigenvector corresponding to a_1, substitute a_1 into A – aI:

$$A - aI = \begin{vmatrix} 1 & -1 \\ 4 & -4 \end{vmatrix}$$

Because $(A - aI)x = 0$, you have

$$\begin{vmatrix} 1 & -1 \\ 4 & -4 \end{vmatrix} \begin{vmatrix} x_1 \\ x_2 \end{vmatrix} = \begin{vmatrix} 0 \\ 0 \end{vmatrix}$$

Because every row of this matrix equation must be true, you know that $x_1 = x_2$. And that means that up to an arbitrary constant, the eigenvector corresponding to a_1 is

$$c\begin{vmatrix} 1 \\ 1 \end{vmatrix}$$

Drop the arbitrary constant and just write this as

$$\begin{vmatrix} 1 \\ 1 \end{vmatrix}$$

How about the eigenvector corresponding to a_2? Plugging a_2 in gives you

$$A - aI = \begin{vmatrix} 4 & -1 \\ 4 & -1 \end{vmatrix}$$

Then you have

$$\begin{vmatrix} 4 & -1 \\ 4 & -1 \end{vmatrix} \begin{vmatrix} x_1 \\ x_2 \end{vmatrix} = \begin{vmatrix} 0 \\ 0 \end{vmatrix}$$

So $4x_1 = x_2$, and that means that up to an arbitrary constant, the eigenvector corresponding to a_2 is

$$c\begin{vmatrix} 1 \\ 4 \end{vmatrix}$$

Drop the arbitrary constant and just write this as

$$\begin{vmatrix} 1 \\ 4 \end{vmatrix}$$

Chapter 2

No Handcuffs Involved: Bound States in Energy Wells

In This Chapter

- Getting wave functions in potential wells
- Solving infinite square wells
- Determining energy levels
- Solving problems where particles impact potential barriers

An *energy well* is formed by a potential (such as an electric potential) that restricts the motion of a particle (which would be a charged particle if you're dealing with an energy well formed by an electric potential). Particles trapped in energy wells are bound there, and they move back and forth.

The motion of such particles is well defined by quantum physics, as are the energy levels. The real value of quantum physics becomes apparent when you work on the microscopic scale, and nowhere is that more true than with energy wells. If you have a macroscopic system, such as a tennis ball stuck in an energy well, the fact that the tennis ball has a spectrum of allowed energy levels isn't apparent. The situation becomes much clearer when you're dealing with, say, an electron in an energy well on the order of an atom in size. On that scale, it immediately becomes clear that the electron is allowed only certain wave functions to describe its motion and certain energy levels.

This chapter is all about solving problems involving energy wells. You can tackle these kinds of problems by finding the allowed wave functions and energy levels.

Starting with the Wave Function

Wave functions are primary to quantum physics. They represent the probability amplitude that a particle will be in a certain state, and they're the solution to the Schrödinger equation. When you have a particle's wave function, you have its energy, expected location, and more.

Square wells are just square potentials (for example, electrical potentials), created to resemble Figure 2-1. They're called square wells because they can trap particles in them, such that the particles bounce off the walls.

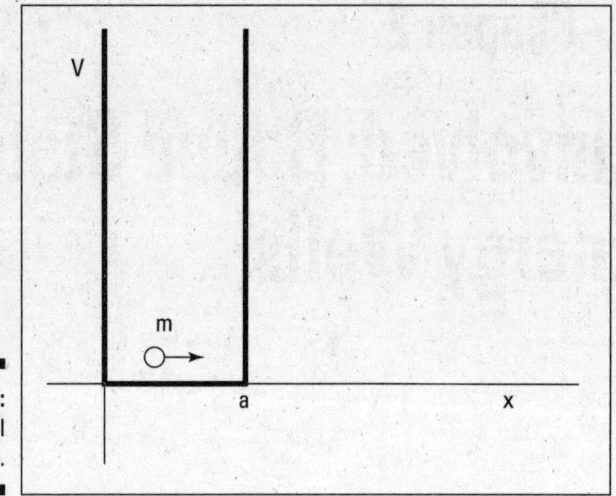

Figure 2-1: A potential well.

In this figure, the energy well's walls are infinite, so you can describe the energy well like this:

$$V(x) = \begin{cases} \infty & x < 0 \\ 0 & 0 \le x \le a \\ \infty & x > a \end{cases}$$

How does a particle trapped in this energy well behave? Quantum physics handles problems like this with the Schrödinger equation, which adds the kinetic and potential energies to give you the total energy in terms of the particle's wave function, like this, where *m* is the mass of the particle:

$$\frac{-\hbar^2}{2m}\Delta\psi(\mathbf{r}) + V(\mathbf{r})\psi(\mathbf{r}) = E\psi(\mathbf{r})$$

Writing out the Laplacian gives you the following:

$$\frac{-\hbar^2}{2m}\left(\frac{\partial^2}{\partial x^2} + \frac{\partial^2}{\partial y^2} + \frac{\partial^2}{\partial z^2}\right)\psi(\mathbf{r}) + V(\mathbf{r})\psi(\mathbf{r}) = E\psi(\mathbf{r})$$

What good is the wave function? The wave function is a particle's *probability amplitude* (the square root of the probability), and you have to square it to get to an actual probability. You can use the probability amplitude to get expectation values, such as the expected value of the *x* position of the particle, which is how you tie wave functions to real, observable quantities:

$$\langle x \rangle = \int_0^a x\left|\psi(x)\right|^2 dx$$

The following example looks at the kinds of solutions you get for the Schrödinger equation for particles trapped in square wells.

Q. What is the wave function for a particle trapped in this one-dimensional well?

$$V(x)=\begin{cases}\infty & x<0\\ 0 & 0\le x\le a\\ \infty & x>a\end{cases}$$

A. $\psi(x)=\mathrm{A}\sin(kx)+\mathrm{B}\cos(kx)$. Here's how to solve this problem:

1. **Write the Schrödinger equation in one dimension.**

   The particle is trapped in a one-dimensional well, so you're interested in only the *x* dimension. You don't need the *y* and *z* terms, so the Schrödinger equation looks like this:

   $$\frac{-\hbar^2}{2m}\frac{d^2}{dx^2}\psi(x)+V(x)\psi(x)=E\psi(x)$$

2. **The particle is trapped in the well, so substitute for V(*x*) for the region inside the well, where $0\le x\le a$.**

   $$\frac{-\hbar^2}{2m}\frac{d^2}{dx^2}\psi(x)=E\psi(x)$$

3. **Find the allowed wave functions by solving the second-order differential equation.**

   The solutions are of the form

   $$\psi_1(x)=\mathrm{A}\sin(kx)$$
   $$\psi_2(x)=\mathrm{B}\cos(kx)$$

   where A, B, and *k* are yet to be determined.

1. Determine *k* in terms of the particle's energy, E, for the solutions $\psi_1(x)$ = A sin(*kx*) and $\psi_2(x)$ = B cos(*kx*), which were just found in the preceding example.

Solve It

2. Show that $\psi(x)$ = A sin(*kx*) + B cos(*kx*) is a solution of the Schrödinger equation for a particle trapped in a square well.

Solve It

Determining Allowed Energy Levels

Figuring out the allowed energy levels (E) is another function you need to know how to do. Because all the systems in this book are quantized, they can only have certain energy levels, called their *allowed energy levels*. Figuring out those energy levels is an important part of understanding what makes those systems tick. The good news is that making this determination isn't too difficult. To do so, you simply match the boundary conditions of the wave function.

A particle bound in this energy well is as follows:

$$V(x) = \begin{cases} \infty & x < 0 \\ 0 & 0 \le x \le a \\ \infty & x > a \end{cases}$$

And the wave function looks like this:

$$\psi(x) = A\sin(kx) + B\cos(kx)$$

where $k = \frac{(2mE)^{1/2}}{\hbar}$ and you still have to determine the values of A and B.

The following example breaks down how you figure out allowed energy levels.

Q. What are the allowed energy levels for a particle trapped in this one-dimensional well?

$$V(x) = \begin{cases} \infty & x < 0 \\ 0 & 0 \le x \le a \\ \infty & x > a \end{cases}$$

A. $\mathbf{E = \frac{n^2\hbar^2\pi^2}{2ma^2} \quad n = 1, 2, 3}$

Because the well is infinite at $x = 0$ and $x = a$, this must be true for the wave equation $\psi(x) = A\sin(kx) + B\cos(kx)$:

- $\psi(0) = 0$
- $\psi(a) = 0$

The fact that $\psi(0) = 0$ tells you that B must be 0, because $\cos(0) = 1$. And the fact that $\psi(a) = 0$ tells you that $\psi(a) = A\sin(ka) = 0$. Because the sine is 0 when its argument is a multiple of π, this means that the following is true, where $n = 1, 2, 3$, and so forth:

$$ka = n\pi \quad n = 1, 2, 3, \ldots$$

This equation becomes

$$k = \frac{n\pi}{a} \quad n = 1, 2, 3, \ldots$$

Finally, rewrite the equation and solve for E. Because $k^2 = \frac{2mE}{\hbar^2}$, you get

$$\frac{2mE}{\hbar^2} = \frac{n^2\pi^2}{a^2} \quad n = 1, 2, 3, \ldots$$

So solve for E, which gives you the following:

$$E = \frac{n^2\hbar^2\pi^2}{2ma^2} \quad n = 1, 2, 3, \ldots$$

Remember: These are quantized energy states, corresponding to the quantum numbers 1, 2, 3, and so on. *Quantized* energy states are the allowed energy states of a system.

3. What is the lowest energy level of a particle of mass *m* trapped in this square well?

$$V(x)=\begin{cases}\infty & x<0\\ 0 & 0\le x\le a\\ \infty & x>a\end{cases}$$

Solve It

4. Assume you have an electron, mass 9.11×10^{-31} kg, confined to an infinite one-dimensional square well of width of the order of *Bohr radius* (the average radius of an electron's orbit in a hydrogen atom, about 5×10^{-11} meters). What is the energy of this electron?

Solve It

Putting the Finishing Touches on the Wave Function by Normalizing It

You know that $\psi(x) = A\sin(kx)$, but how do you solve for A? You can solve for A because the wave function has to be *normalized* over all space; that is, the square of the wave function, which is the probability that the particle will be in a certain location, has to equal 1 when integrated over all space.

For a square well like this

$$V(x) = \begin{cases} \infty & x < 0 \\ 0 & 0 \le x \le a \\ \infty & x > a \end{cases}$$

you have the following wave equation:

$$\psi(x) = A\sin\left(\frac{n\pi x}{a}\right)$$

So you want to solve for A. The following example shows you how to do so, and the practice problems help you get more familiar with the concept.

Q. Determine the normalization condition that the wave function solution to this square well must satisfy:

$$V(x) = \begin{cases} \infty & x < 0 \\ 0 & 0 \le x \le a \\ \infty & x > a \end{cases}$$

A. $1 = \int_0^a |\psi(x)|^2\, dx$. Here's how to solve the problem:

1. **Set up the square of the wave function for the interval from – to and set it equal to 1.**

   Wave functions have to be normalized — that is, the probability of finding the particle between x and dx, $|\psi(x)|^2\, dx$, must add up to 1 when you integrate over all space:

   $$1 = \int_{-\infty}^{\infty} |\psi(x)|^2\, dx$$

2. **Set the integral's boundaries.**

   The square well looks like this:

   $$V(x) = \begin{cases} \infty & x < 0 \\ 0 & 0 \le x \le a \\ \infty & x > a \end{cases}$$

   which means that $\psi(x) = 0$ outside the region $0 < x < a$, so the normalization condition becomes this:

   $$1 = \int_0^a |\psi(x)|^2\, dx$$

5. Use the following normalization condition to solve for A if $\psi(x) = A\sin(n\pi x/a)$:

$$1 = \int_0^a |\psi(x)|^2 \, dx$$

Solve It

6. Using the form of A you solved for in the preceding problem, find the complete form of $\psi(x)$.

Solve It

Translating to a Symmetric Square Well

The region inside a square well doesn't have to go from 0 to a, as it does in the preceding sections. You can shift the well to the left so that it's symmetric around the origin instead. In order to shift, you move the square well so that it extends from $-a/2$ to $a/2$. The new infinite square well looks like this:

$$V(x)=\begin{cases}\infty & x<\frac{-a}{2}\\ 0 & \frac{-a}{2}\le x\le\frac{a}{2}\\ \infty & x>\frac{a}{2}\end{cases}$$

Often, you see problems with square wells that are symmetric around the origin like this, so it pays to know how to shift between square well potentials centered around the origin and those that start at the origin.

Q. Given that the following square well

$$V(x)=\begin{cases}\infty & x<0\\ 0 & 0\le x\le a\\ \infty & x>a\end{cases}$$

creates wave functions like this:

$$\psi(x)=\left(\frac{2}{a}\right)^{1/2}\sin\left(\frac{n\pi x}{a}\right)\quad n=1,\ 2,\ 3,\ \ldots$$

what are the wave functions when you shift the square well to the following coordinates?

$$V(x)=\begin{cases}\infty & x<\frac{-a}{2}\\ 0 & \frac{-a}{2}\le x\le\frac{a}{2}\\ \infty & x>\frac{a}{2}\end{cases}$$

A. $\boldsymbol{\psi(x)=\left(\frac{2}{a}\right)^{1/2}\sin\left(\frac{n\pi}{a}\left(x+\frac{a}{2}\right)\right)\quad n=1,\ 2,\ 3,\ \ldots}$

To shift the sine function along the x-axis, you subtract the amount of the shift from x. You want to move the wave function $\frac{-a}{2}$ units, so you can translate from this old square well to the new one by adding $\frac{a}{2}$ to x — that is, $x-\left(\frac{-a}{2}\right)$. So if the original wave function is

$$\psi(x)=\left(\frac{2}{a}\right)^{1/2}\sin\left(\frac{n\pi}{a}x\right)\quad n=1,\ 2,\ 3,\ \ldots$$

then you can write the wave function for the new square well like this:

$$\psi(x)=\left(\frac{2}{a}\right)^{1/2}\sin\left(\frac{n\pi}{a}\left(x+\frac{a}{2}\right)\right)\quad n=1,\ 2,\ 3,\ \ldots$$

7. Simplify the new $\psi(x)$ in terms of sines and cosines, without the $(x + \frac{a}{2})$ term:

$$\psi(x) = \left(\frac{2}{a}\right)^{1/2} \sin\left(\frac{n\pi}{a}\left(x + \frac{a}{2}\right)\right) \quad n = 1, 2, 3, \ldots$$

Solve It

8. What are the first four wave functions for a particle in this symmetric square well?

$$V(x) = \begin{cases} \infty & x < \frac{-a}{2} \\ 0 & \frac{-a}{2} \le x \le \frac{a}{2} \\ \infty & x > \frac{a}{2} \end{cases}$$

Solve It

Banging into the Wall: Step Barriers When the Particle Has Plenty of Energy

You often see problems involving potential steps. *Potential steps* are just like what they sound like — the potential changes from one level to another. Check out Figure 2-2 for an example of a potential step.

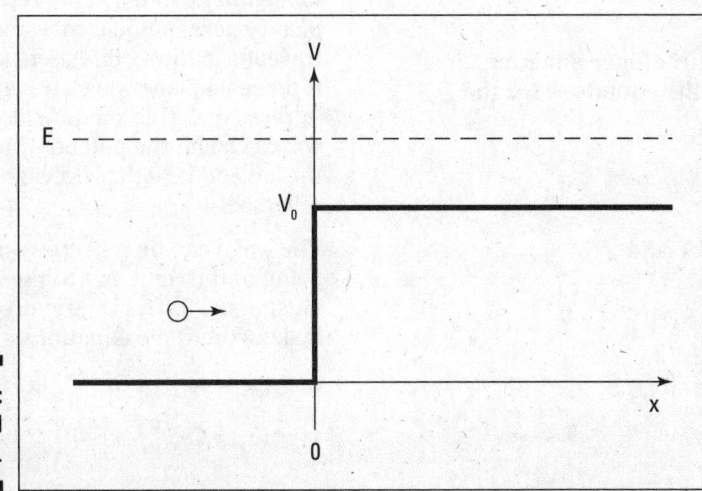

Figure 2-2: A potential step.

The potential for Figure 2-2 looks like this:

$$V(x)=\begin{cases}0 & x<0\\ V_0 & x>0\end{cases}$$

A particle traveling to the right that has energy, E, that's greater than V_0 can easily enter the region $x > 0$ because it has enough energy to get past the barrier. (The next section addresses the other case, where $E < V_0$.) However, as you find out in the example problem, not all particles get through.

In the case where $E > V_0$, here's what the Schrödinger equation looks like for the region $x < 0$:

$$\frac{d^2\psi_1}{dx^2}(x)+k_1^2\psi_1(x)=0 \quad x<0$$

where $k_1^2 = 2mE/\hbar^2$. And for the region $x > 0$, this is what the Schrödinger equation looks like:

$$\frac{d^2\psi_2}{dx^2}(x)+k_2^2\psi_2(x)=0 \quad x>0$$

where $k_2^2 = 2m(E-V_0)/\hbar^2$.

It turns out that you can calculate R, the fraction of the wave function that's *reflected,* and T, the fraction of the wave function that's *transmitted.* Check out the following example for solving the Schrödinger equation with this case, and then try the practice problems.

Q. Solve the Schrödinger equation for the regions $x < 0$ and $x > 0$.

A. **The answer is**

- $\psi_1(x)=Ae^{ik_1x}+Be^{-ik_1x} \quad x<0$
- $\psi_2(x)=Ce^{ik_2x} \quad x>0$

where $k_1^2 = 2mE/\hbar^2$ **and** $k_2^2 = 2m(E-V_0)/\hbar^2$. Here's how you solve it:

1. **Start with the Schrödinger equation, which gives you the equations for the wave functions.**

$$\frac{d^2\psi_1}{dx^2}(x)+k_1^2\psi_1(x)=0 \quad x<0$$

where $k_1^2 = 2mE/\hbar^2$ and

$$\frac{d^2\psi_2}{dx^2}(x)+k_2^2\psi_2(x)=0 \quad x>0$$

where $k_2^2 = 2m(E-V_0)/\hbar^2$.

2. **Solve for ψ.**

To do so, treat these equations as simple second-order differential equations, which gives you these solutions:

- $\psi_1(x)=Ae^{ik_1x}+Be^{-ik_1x} \quad x<0$
- $\psi_2(x)=Ce^{ik_2x}+De^{-ik_2x} \quad x>0$

where A, B, C, and D are constants.

Remember: Note that e^{ikx} represents *plane waves* (unfocused wave funtions) traveling in the $+x$ direction, and e^{-ikx} represents plane waves traveling in the $-x$ direction. This solution means that waves can hit the potential step from the left and be either transmitted or reflected.

The wave can be reflected only if it's going to the right, not to the left, so D must equal 0. The D term drops out, making the wave equations

- $\psi_1(x)=Ae^{ik_1x}+Be^{-ik_1x} \quad x<0$
- $\psi_2(x)=Ce^{ik_2x} \quad x>0$

9. Calculate the reflection coefficient, R, and transmission coefficient, T, for the wave where

- $\psi_1(x) = Ae^{ik_1x} + Be^{-ik_1x} \quad x < 0$
- $\psi_2(x) = Ce^{ik_2x} \quad x > 0$

in terms of A, B, and C, where R is the probability that the particle will be reflected and T is the probability the wave will be transmitted through the potential step.

Solve It

10. Solve for the reflection coefficient R and the transmission coefficient T for this wave function:

- $\psi_1(x) = Ae^{ik_1x} + Be^{-ik_1x} \quad x < 0$
- $\psi_2(x) = Ce^{ik_2x} \quad x > 0$

in terms of k_1 and k_2, where $k_1^2 = 2mE/\hbar^2$ and $k_2^2 = 2m(E - V_0)/\hbar^2$.

Solve It

Hitting the Wall: Step Barriers When the Particle Doesn't Have Enough Energy

When dealing with potential steps, you may encounter another case: when E, the energy of the particle, is less than V_0, the potential of the step. That doesn't make a difference in the region $x < 0$, before the step — the solution to the Schrödinger equation stays the same, regardless of how much energy the particle has. Here's what the solution looks like for $x < 0$ (see the preceding section for info on how to find this answer):

$$\psi_1(x) = Ae^{ik_1x} + Be^{-ik_1x} \quad x < 0$$

But in this case, the particle doesn't have enough energy to make it into the region $x > 0$, according to classical physics. However, it can actually get there according to quantum physics. Check out the following example to figure out how to solve for the region $x > 0$. Then try the practice problems.

Q. Solve the Schrödinger equation for a potential step where the particle has energy, E, that's less than the potential of the step, V_0:

$$V(x) = \begin{cases} 0 & x < 0 \\ V_0 & x > 0 \end{cases}$$

A. **Here are the answers:**

- $\psi_i(x) = Ae^{ik_1x}$
- $\psi_r(x) = Be^{-ik_1x}$
- $\psi_t(x) = Ce^{-k_2x}$

where $k_1^2 = 2mE/\hbar^2$ **and** $k_2^2 = 2m(V_0 - E)/\hbar^2$. Follow these steps to solve the problem:

1. **Start with the Schrödinger equation for $x > 0$.**

   $$\frac{d^2\psi_2}{dx^2}(x) + k^2\psi_2(x) = 0 \quad x > 0$$

   where $k^2 = 2m(E - V_0)/\hbar^2$. $E - V_0$ is less than 0, which would make k imaginary. That's physically impossible, so change the sign in the Schrödinger equation from plus to minus:

   $$\frac{d^2\psi_2}{dx^2}(x) - k^2\psi_2(x) = 0 \quad x > 0$$

   and use this for k_2 (note that this is positive if $E < V_0$):

   $$k_2^2 = 2m(V_0 - E)/\hbar^2$$

2. **Solve the differential equation.**

   By doing so, you find two linearly independent solutions:

   - $\psi(x) = Ce^{-k_2x}$
   - $\psi(x) = De^{k_2x}$

   And the general solution to the Schrödinger equation is

   $$\psi_2(x) = Ce^{-k_2x} + De^{k_2x} \quad x > 0$$

   Remember: Wave functions have to be finite everywhere, and the second term is clearly not finite as x approaches infinity, so D must equal 0, which makes the solution for $x > 0$ equal to

   $$\psi_2(x) = Ce^{-k_2x} \quad x > 0$$

   So the wave functions for the two regions are as follows:

   - $\psi_1(x) = Ae^{ik_1x} + Be^{-ik_1x} \quad x < 0$
   - $\psi_2(x) = Ce^{-k_2x} + Be^{-ik_1x} \quad x > 0$

   You can also cast this wave function in terms of the incident, reflected, and transmitted wave functions, $\psi_i(x)$, $\psi_r(x)$ and $\psi_t(x)$. When you do, you get the following:

   - $\psi_i(x) = Ae^{ik_1x}$
   - $\psi_r(x) = Be^{-ik_1x}$
   - $\psi_t(x) = Ce^{-k_2x}$

11. Calculate the reflection coefficient, R, and transmission coefficient, T, for the wave function of a particle encountering a potential step, V_0

$$V(x) = \begin{cases} 0 & x < 0 \\ V_0 & x > 0 \end{cases}$$

where the energy of the particle $E < V_0$ and the wave function is

- $\psi_i(x) = Ae^{ik_1x}$
- $\psi_r(x) = Be^{-ik_1x}$
- $\psi_t(x) = Ce^{-k_2x}$

where $k_1^2 = 2m\frac{E}{\hbar^2}$ and $k_2^2 = 2m\left(\frac{V_0 - E}{\hbar^2}\right)$.

Solve It

12. Calculate the probability density, $P(x) = |\psi_t(x)|^2$, for the particle from the preceding problem to be in the region $x > 0$. *Probability density,* when integrated over all space, gives you a probability of 1 that the particle will be found somewhere in all space. The probability that the particle will be found in the region dx is $|\psi_t(x)|^2\,dx$.

Solve It

Plowing through a Potential Barrier

Sometimes the particles have to try to plow through a potential that represents a barrier. A *barrier* is like a step, but whereas the barrier also returns to its original value, a step doesn't. A potential barrier looks like this:

$$V(x)=\begin{cases}\infty & x<0\\ 0 & 0\le x\le a\\ \infty & x>a\end{cases}$$

You can see what this potential barrier looks like in Figure 2-3.

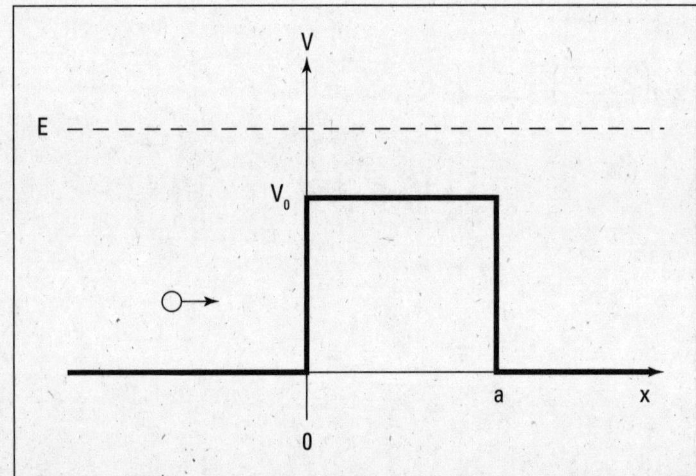

Figure 2-3: A potential barrier.

To solve the Schrödinger equation for a potential barrier, you have to consider two cases based on whether the particle has more or less energy than the potential barrier. That is, if E is the energy of the incident particle, the two cases to consider are

- $E > V_0$
- $E < V_0$

Q. Find the wave function (up to arbitrary normalization coefficients) for a particle encountering the following potential barrier where the energy of the particle $E > V_0$:

$$V(x) = \begin{cases} 0 & x < 0 \\ V_0 & 0 \le x \le a \\ 0 & x > a \end{cases}$$

A. **Here are the answers:**

- $\psi_1(x) = Ae^{ik_1x} + Be^{-ik_1x} \quad x < 0$
- $\psi_2(x) = Ce^{ik_2x} + De^{-ik_2x} \quad 0 \le x \le a$
- $\psi_3(x) = Ee^{ik_1x} \quad x > a$

where $k_1^2 = \frac{2mE}{\hbar^2}$ **and** $k_2^2 = \frac{2m(E - V_0)}{\hbar^2}$.

Here's how to find the solution:

1. **Write the Schrödinger equation.**

   Here's what the equation looks like for $x < 0$:

   $$\frac{d^2\psi_1}{dx^2}(x) + k_1^2\psi_1(x) = 0 \quad x < 0$$

   where $k_1^2 = \frac{2mE}{\hbar^2}$.

   For the region $0 \le x \le a$, this is what the Schrödinger equation looks like:

   $$\frac{d^2\psi_2}{dx^2}(x) + k_2^2\psi_2(x) = 0 \quad 0 \le x \le a$$

   where $k_2^2 = \frac{2m(E - V_0)}{\hbar^2}$.

   For the region $x > a$, this is what the Schrödinger equation looks like:

   $$\frac{d^2\psi_3}{dx^2}(x) + k_1^2\psi_3(x) = 0 \quad x > a$$

   where $k_1^2 = \frac{2mE}{\hbar^2}$.

2. **Solve these differential equations for $\psi_1(x)$, $\psi_2(x)$, and $\psi_3(x)$.**

   Doing so gives you the following:

   - $\psi_1(x) = Ae^{ik_1x} + Be^{-ik_1x} \quad x < 0$
   - $\psi_2(x) = Ce^{ik_2x} + De^{-ik_2x} \quad 0 \le x \le a$
   - $\psi_3(x) = Ee^{ik_1x} + Fe^{-ik_1x} \quad x > a$

3. **Set F to 0.**

   Because there's no leftward-traveling wave in the $x > a$ region, $F = 0$, so

   $$\psi_3(x) = Ee^{ik_1x} \quad x > a$$

13. Calculate the transmission coefficient, T, for the wave function where $E > V_0$ and

$$V(x) = \begin{cases} 0 & x < 0 \\ V_0 & 0 \le x \le a \\ 0 & x > a \end{cases}$$

and where

- $\psi_1(x) = Ae^{ik_1x} + Be^{-ik_1x} \quad x < 0$
- $\psi_2(x) = Ce^{ik_2x} + De^{-ik_2x} \quad 0 \le x \le a$
- $\psi_3(x) = Ee^{ik_1x} \quad x > a$

where $k_1^2 = \frac{2mE}{\hbar^2}$ and $k_2^2 = \frac{2m(E - V_0)}{\hbar^2}$.

Solve It

14. Calculate the reflection coefficient, R, for the wave function where $E > V_0$ and

$$V(x) = \begin{cases} 0 & x < 0 \\ V_0 & 0 \le x \le a \\ 0 & x > a \end{cases}$$

and where

- $\psi_1(x) = Ae^{ik_1x} + Be^{-ik_1x} \quad x < 0$
- $\psi_2(x) = Ce^{ik_2x} + De^{-ik_2x} \quad 0 \le x \le a$
- $\psi_3(x) = Ee^{ik_1x} \quad x > a$

where $k_1^2 = \frac{2mE}{\hbar^2}$ and $k_2^2 = \frac{2m(E - V_0)}{\hbar^2}$.

Solve It

15. Find the wave function (up to arbitrary normalization coefficients) for a particle encountering the following potential barrier where the energy of the particle is $E < V_0$ (that is, the particle has less energy than the potential barrier this time):

$$V(x)=\begin{cases} 0 & x<0 \\ V_0 & 0\le x\le a \\ 0 & x>a \end{cases}$$

Solve It

16. Calculate the reflection and transmission coefficients for a particle hitting a potential barrier of V_0 where the particle's energy $E < V_0$ and

$$V(x)=\begin{cases} 0 & x<0 \\ V_0 & 0\le x\le a \\ 0 & x>a \end{cases}$$

and where

- $\psi_1(x)=Ae^{ik_1x}+Be^{-ik_1x} \quad x<0$
- $\psi_2(x)=Ce^{ik_2x}+De^{-ik_2x} \quad 0\le x\le a$
- $\psi_3(x)=Ee^{ik_1x} \quad x>a$

where $k_1^2=\frac{2mE}{\hbar^2}$ and $k_2^2=\frac{2m(E-V_0)}{\hbar^2}$.

Solve It

Answers to Problems on Bound States

The following are the answers to the practice questions presented earlier in this chapter. I repeat the questions, give the answers in bold, and show you how to work out the answers, step by step.

1. Determine k in terms of the particle's energy, E, for the solutions $\psi_1(x) = A \sin(kx)$ and $\psi_2(x) = B \cos(kx)$ that you just found in the preceding example. **Here's the answer:**

$$\boldsymbol{k = \frac{(2m\mathrm{E})^{1/2}}{\hbar}}$$

To solve, start with the Schrödinger equation for a particle in the square well shown in Figure 2-1. This is a one-dimensional well, so you don't need the y and z coordinates:

$$\frac{-\hbar^2}{2m}\frac{d^2}{dx^2}\psi(x) + \mathrm{V}(x)\psi(x) = \mathrm{E}\psi(x)$$

Because $\mathrm{V}(x) = 0$ inside the well, the second term drops out:

$$\frac{-\hbar^2}{2m}\frac{d^2}{dx^2}\psi(x) = \mathrm{E}\psi(x)$$

Write this equation in terms of k as

$$\frac{d^2}{dx^2}\psi(x) + k^2\psi(x) = 0$$

where $k^2 = \frac{2m\mathrm{E}}{\hbar^2}$.

Now you have a second-order differential equation. When you solve the equation, you come up with the following:

- $\psi_1(x) = A \sin(kx)$
- $\psi_2(x) = B \cos(kx)$

where A and B are constants that are yet to be determined and

$$k^2 = \frac{(2m\mathrm{E})^{1/2}}{\hbar}$$

Here's the final wave function:

$$\psi(x) = A \sin(kx) + B \cos(kx)$$

2. Show that $\psi(x) = A \sin(kx) + B \cos(kx)$ is a solution of the Schrödinger equation for a particle trapped in a square well. **The equation $\psi(x) = A \sin(kx) + B \cos(kx)$ works as a solution.**

Write the Schrödinger equation for a particle in the square well shown in Figure 2-1:

$$\frac{-\hbar^2}{2m}\frac{d^2}{dx^2}\psi(x) + \mathrm{V}(x)\psi(x) = \mathrm{E}\psi(x)$$

Because $\mathrm{V}(x) = 0$ inside the well, the second term equals zero, so the equation becomes

$$\frac{-\hbar^2}{2m}\frac{d^2}{dx^2}\psi(x) = \mathrm{E}\psi(x)$$

You can write this as

$$\frac{d^2}{dx^2}\psi(x)+k^2\psi(x)=0$$

where $k^2=\frac{2mE}{\hbar^2}$.

Plugging in $\psi(x)$ = A sin(kx) + B cos(kx) gives you the following:

$$\frac{d^2}{dx^2}\left(A\sin(kx)+B\cos(kx)\right)+k^2\left(A\sin(kx)+B\cos(kx)\right)=0$$

Simplifying, this becomes

$$-k^2\left(A\sin(kx)+B\cos(kx)\right)+k^2\left(A\sin(kx)+B\cos(kx)\right)=0$$

which does equal 0, as the equation requires.

3 What is the lowest energy level of a particle of mass m trapped in this square well?

$$V(x)=\begin{cases}\infty & x<0\\ 0 & 0\le x\le a\\ \infty & x>a\end{cases}$$

The answer is

$$\mathbf{E=\frac{\hbar^2\pi^2}{2ma^2}}$$

The energy levels for a particle trapped in a square well are

$$E=\frac{n^2\hbar^2\pi^2}{2ma^2}\quad n=1,\ 2,\ 3,\ \ldots$$

Although n = 0 is technically a solution to the Schrödinger equation, it yields $\psi(0)$ = 0, so it's not a physical solution — the physical solutions begin with n = 1. The first physical state corresponds to n = 1, which gives you the following, because $1^2 = 1$:

$$E=\frac{\hbar^2\pi^2}{2ma^2}$$

This is the *ground state,* the lowest allowed energy state that the particle can occupy.

4 Assume you have an electron, mass 9.11×10^{-31} kg, confined to an infinite one-dimensional square well of width of the order of *Bohr radius* (the average radius of an electron's orbit in a hydrogen atom, about 5×10^{-11} meters). What is the energy of this electron? **About 16 electron volts.**

The energy levels for a particle trapped in a square well are

$$E=\frac{n^2\hbar^2\pi^2}{2ma^2}\quad n=1,\ 2,\ 3,\ \ldots$$

The ground energy state is

$$E=\frac{\hbar^2\pi^2}{2ma^2}$$

Now just plug in the physical constants, which gives you the following:

$$E \approx \frac{\left(1.05\times10^{-34}\right)^2 \left(3.14\right)^2}{2\left(9.11\times10^{-31}\right)\left(5\times10^{-11}\right)^2}$$

Perform the multiplication to find the energy in joules:

$$E \approx \frac{\left(1.05\times10^{-34}\right)^2 \left(3.14\right)^2}{2\left(9.11\times10^{-31}\right)\left(5\times10^{-11}\right)^2} = 2.40\times10^{-18}\text{ joules}$$

That's about 16 *electron volts* (eV — the amount of energy one electron gains falling through 1 volt; 1 eV = 1.6×10^{-19} joules).

This provides a good estimate for the energy of an electron in the ground state of a hydrogen atom (13.6 eV), so you can say you're certainly in the right quantum ballpark.

5 Use the following normalization condition to solve for A if $\psi(x) = A\sin(n\pi x/a)$:

$$1 = \int_0^a \left|\psi(x)\right|^2 dx$$

The answer is $\mathbf{A} = \left(\frac{2}{a}\right)^{1/2}$**.**

Start with the normalization condition of the wave function $\psi(x) = A\sin(n\pi x/a)$, which is

$$= \int_0^a \left|\psi(x)\right|^2 dx$$

Now plug $\psi(x) = A\sin(n\pi x/a)$ into the normalization condition:

$$1 = \left|A\right|^2 \int_0^a \sin^2\left(\frac{n\pi x}{a}\right) dx$$

Find the value of the integral:

$$\int_0^a \sin^2\left(\frac{n\pi x}{a}\right) dx = \frac{a}{2}$$

Replace the integral with $\frac{a}{2}$ in the normalization equation, which gives you the following:

$$1 = \left|A\right|^2 \left(\frac{a}{2}\right)$$

Finally, solve for A:

$$A = \left(\frac{2}{a}\right)^{1/2}$$

6 Using the form of A you solved for in the preceding problem, find the complete form of $\psi(x)$.

$$\boldsymbol{\psi(x) = \left(\frac{2}{a}\right)^{1/2} \sin\left(\frac{n\pi x}{a}\right) \quad n = 1,\ 2,\ 3,\ \ldots}$$

The wave function $\psi(x)$ has this form:

$$\psi(x) = A\sin\left(\frac{n\pi x}{a}\right) \quad n = 1,\ 2,\ 3,\ \ldots$$

The solution for problem 5 tells you the value of A:

$$A = \left(\frac{2}{a}\right)^{1/2}$$

Simply plug A into $\psi(x)$ to get the following:

$$\psi(x) = \left(\frac{2}{a}\right)^{1/2} \sin\left(\frac{n\pi x}{a}\right) \quad n = 1, 2, 3, \ldots$$

7 Simplify the new $\psi(x)$ in terms of sines and cosines, without the $\left(x + \frac{a}{2}\right)$ term:

$$\psi(x) = \left(\frac{2}{a}\right)^{1/2} \sin\left(\frac{n\pi}{a}\left(x + \frac{a}{2}\right)\right) \quad n = 1, 2, 3, \ldots$$

The answer is

$$\psi(x) = \left(\frac{2}{a}\right)^{1/2} \cos\left(\frac{n\pi}{a}(x)\right) \quad n = 1, 3, 5, \ldots$$
$$+\left(\frac{2}{a}\right)^{1/2} \sin\left(\frac{n\pi}{a}(x)\right) \quad n = 2, 4, 6, \ldots$$

Start with this wave function:

$$\psi(x) = \left(\frac{2}{a}\right)^{1/2} \sin\left(\frac{n\pi}{a}\left(x + \frac{a}{2}\right)\right) \quad n = 1, 2, 3, \ldots$$

Using some trigonometry gives you

$$\psi(x) = \left(\frac{2}{a}\right)^{1/2} \cos\left(\frac{n\pi}{a}(x)\right) \quad n = 1, 3, 5, \ldots$$
$$+\left(\frac{2}{a}\right)^{1/2} \sin\left(\frac{n\pi}{a}(x)\right) \quad n = 2, 4, 6, \ldots$$

In other words, the result is a mix of sines and cosines.

8 What are the first four wave functions for a particle in this symmetric square well?

$$V(x) = \begin{cases} \infty & x < \frac{-a}{2} \\ 0 & \frac{-a}{2} \leq x \leq \frac{a}{2} \\ \infty & x > \frac{a}{2} \end{cases}$$

Here are the answers:

- $\psi_1(x) = \left(\frac{2}{a}\right)^{1/2} \cos\left(\frac{\pi x}{a}\right)$
- $\psi_2(x) = \left(\frac{2}{a}\right)^{1/2} \sin\left(\frac{2\pi x}{a}\right)$
- $\psi_3(x) = \left(\frac{2}{a}\right)^{1/2} \cos\left(\frac{3\pi x}{a}\right)$
- $\psi_4(x) = \left(\frac{2}{a}\right)^{1/2} \sin\left(\frac{4\pi x}{a}\right)$

For this square well

$$V(x)=\begin{cases}\infty & x<-\frac{a}{2}\\ 0 & -a\le x\le\frac{a}{2}\\ \infty & x>\frac{a}{2}\end{cases}$$

Here's the wave function (see the preceding problem for details):

$$\psi(x)=\left(\frac{2}{a}\right)^{1/2}\cos\left(\frac{n\pi}{a}(x)\right)\quad n=1,\ 3,\ 5,\ \ldots$$

$$\psi(x)=\left(\frac{2}{a}\right)^{1/2}\sin\left(\frac{n\pi}{a}(x)\right)\quad n=2,\ 4,\ 6,\ \ldots$$

To find the first four wave functions, plug in 1 through 4 for *n*. So the first four wave functions are

- $\psi_1(x)=\left(\frac{2}{a}\right)^{1/2}\cos\left(\frac{\pi x}{a}\right)$
- $\psi_2(x)=\left(\frac{2}{a}\right)^{1/2}\sin\left(\frac{2\pi x}{a}\right)$
- $\psi_3(x)=\left(\frac{2}{a}\right)^{1/2}\cos\left(\frac{3\pi x}{a}\right)$
- $\psi_4(x)=\left(\frac{2}{a}\right)^{1/2}\sin\left(\frac{4\pi x}{a}\right)$

Note that the cosines are symmetric around the origin ($\psi(x) = \psi(-x)$) and the sines are anti-symmetric ($-\psi(x) = \psi(-x)$).

9 Calculate the reflection coefficient, R, and transmission coefficient, T, for the wave where

- $\psi_1(x)=\text{A}e^{ik_1x}+\text{B}e^{-ik_1x}\quad x<0$
- $\psi_2(x)=\text{C}e^{ik_2x}\quad x>0$

in terms of A, B, and C, where R is the probability that the particle will be reflected and T is the probability the wave will be transmitted through the potential step. **Here's the answer:**

- $\mathbf{R}=\frac{|\mathbf{B}|^2}{|\mathbf{A}|^2}$
- $\mathbf{T}=\left(\frac{\mathbf{k}_2}{\mathbf{k}_1}\right)\frac{|\mathbf{C}|^2}{|\mathbf{A}|^2}$

Start with this wave function:

- $\psi_1(x) = Ae^{ik_1x} + Be^{-ik_1x} \quad x < 0$
- $\psi_2(x) = Ce^{ik_2x} \quad x > 0$

If J_r is the reflected *current density* (the amount of probability that flows per second per cross-sectional area), J_i is the incident current density, and J_t is the transmitted current density, then R, the reflection coefficient, is

$$R = \frac{J_r}{J_i}$$

And T, the transmission coefficient, is

$$T = \frac{J_t}{J_i}$$

For both the transmitted and the reflection coefficient, you need to know J_i. What is J_i? Because the incident part of the wave is $\psi_i(x) = Ae^{ik_1x}$, the incident current density is the following, where the asterisks denote the complex conjugate:

$$J_i = \frac{i\hbar}{2m}\left[\psi_i(x)\frac{d\psi_i^*(x)}{dx} - \psi_i^*(x)\frac{d\psi_i(x)}{dx}\right]$$

And substituting, this just equals

$$J_i = \frac{i\hbar}{2m}\left[\psi_i(x)\frac{d\psi_i^*(x)}{dx} - \psi_i^*(x)\frac{d\psi_i(x)}{dx}\right] = \frac{\hbar k_1}{m}|A|^2$$

In the same way, you can find J_r and J_t in terms of B and C:

- $J_r = \frac{\hbar k_1}{m} \quad |B|^2$
- $J_t = \frac{\hbar k_2}{m} \quad |C|^2$

So substituting for J_r and J_i, here's what you get for the reflection coefficient:

$$R = \frac{J_r}{J_i} = \frac{|B|^2}{|A|^2}$$

And substituting for J_t and J_i, you find that T, the transmission coefficient, is

$$T = \frac{J_t}{J_i} = \left(\frac{k_2}{k_1}\right)\frac{|C|^2}{|A|^2}$$

10 Solve for the reflection coefficient R and the transmission coefficient T for this wave function:

- $\psi_1(x) = Ae^{ik_1x} + Be^{-ik_1x} \quad x < 0$
- $\psi_2(x) = Ce^{ik_2x} \quad x > 0$

in terms of k_1 and k_2, where $k_1^2 = 2mE/\hbar^2$ and $k_2^2 = 2m(E - V_0)/\hbar^2$.

Here are the answers:

- $\mathbf{R} = \frac{(k_1 - k_2)^2}{(k_1 + k_2)^2}$
- $\mathbf{T} = \frac{4k_1 k_2}{(k_1 + k_2)^2}$

where $k_1^2 = 2m\mathrm{E}/\hbar^2$ and $k_2^2 = 2m(\mathrm{E} - \mathrm{V}_0)/\hbar^2$.

So how do you calculate A, B, and C in terms of k_1 and k_2? You do that with boundary conditions. The boundary conditions here are that $\psi(x)$ and $d\psi(x)/dx$ are continuous across the potential step's boundary. In other words

$$\psi_1(0) = \psi_2(0)$$

and

$$\frac{d\psi_1}{dx}(0) = \frac{d\psi_2}{dx}(0)$$

You know that

- $\psi_1(x) = \mathrm{A}e^{ik_1x} + \mathrm{B}e^{-ik_1x} \quad x < 0$
- $\psi_2(x) = \mathrm{C}e^{ik_2x} \quad x > 0$

If you plug in the boundary conditions, here's what you get:

- $\mathrm{A} + \mathrm{B} = \mathrm{C}$
- $k_1\mathrm{A} - k_1\mathrm{B} = k_2\mathrm{C}$

Now solve for B in terms of A. That gives you the following result:

$$\mathrm{B} = \left(\frac{k_1 - k_2}{k_1 + k_2}\right)\mathrm{A}$$

Solve for C in terms of A, which gives you

$$\mathrm{C} = \left(\frac{2k_1}{k_1 + k_2}\right)\mathrm{A}$$

You don't need to solve for A, because it'll drop out of the ratios for the reflection and transmission coefficients, R and T. $|\mathrm{A}|^2$ appears in the denominators, and when you square $|\mathrm{B}|$ and $|\mathrm{C}|$ in the numerators, the A^2 cancels out:

- $\mathrm{R} = \frac{|\mathrm{B}|^2}{|\mathrm{A}|^2}$
- $\mathrm{T} = \left(\frac{k_1}{k_2}\right)\frac{|\mathrm{C}|^2}{|\mathrm{A}|^2}$

Therefore, you get the following:

- $R = \frac{(k_1 - k_2)^2}{(k_1 + k_2)^2}$
- $T = \frac{4k_1k_2}{(k_1 + k_2)^2}$

11 Calculate the reflection coefficient, R, and transmission coefficient, T, for the wave function of a particle encountering a potential step, V_0

$$V(x) = \begin{cases} 0 & x < 0 \\ V_0 & x > 0 \end{cases}$$

where the energy of the particle $E < V_0$ and the wave function is

- $\psi_i(x) = Ae^{ik_1x}$
- $\psi_r(x) = Be^{-ik_1x}$
- $\psi_t(x) = Ce^{-k_2x}$

where $k_1^2 = 2m\frac{E}{\hbar^2}$ and $k_2^2 = 2m\left(\frac{V_0 - E}{\hbar^2}\right)$. **The answer is R = 1 and T = 0.**

Here are the equations for R and T, where J is the probability current density of the particle:

- $R = \frac{J_r}{J_i}$
- $T = \frac{J_t}{J_i}$

Here's what J_t looks like:

$$J_t = \frac{i\hbar}{2m}\left[\psi_t(x)\frac{d\psi_t^*(x)}{dx} - \psi_t^*(x)\frac{d\psi_t(x)}{dx}\right]$$

But because $\psi_t(x) = Ce^{-ik_2x}$ is completely real (which means that you can remove the * symbols because they indicate a complex conjugate), you get the following:

$$J_t = \frac{i\hbar}{2m}\left[\psi_t(x)\frac{d\psi_t(x)}{dx} - \psi_t(x)\frac{d\psi_t(x)}{dx}\right]$$

And doing the subtraction shows you that this, of course, is zero:

$$J_t = \frac{i\hbar}{2m}\left[\psi_t(x)\frac{d\psi_t(x)}{dx} - \psi_t(x)\frac{d\psi_t(x)}{dx}\right] = 0$$

$J_t = 0$, so $T = 0$. And if $T = 0$, then $R = 1$. That means that there is complete reflection.

12 Calculate the probability density, $P(x) = |\psi_t(x)|^2$, for the particle from the preceding problem to be in the region $x > 0$. *Probability density,* when integrated over all space, gives you a probability of 1 that the particle will be found somewhere in all space. The probability that the particle will be found in the region dx is $|\psi_t(x)|^2\, dx$. **If the particle's wave function is**

- $\psi_i(x) = Ae^{ik_1x}$
- $\psi_r(x) = Be^{ik_1x}$
- $\psi_t(x) = Ce^{-k_2x}$

where $k_1^2 = 2mE/\hbar^2$ **and** $k_2^2 = 2m(E - V_0)/\hbar^2$**, the probability density for** $x > 0$ **is**

$$P(x) = \frac{4k_1^2|A|^2 e^{-2k_2x}}{k_1^2 + k_2^2}$$

The probability density for $x > 0$ is

$$P(x) = |\psi_t(x)|^2$$

Plug in $\psi_t(x)$ to get

$$P(x) = |\psi_t(x)|^2 = |C|^2 e^{-2k_2x}$$

Now use the continuity conditions — $\psi_1(0) = \psi_2(0)$ and $\frac{d\psi_1}{dx}(0) = \frac{d\psi_2}{dx}(0)$ — and solve for C in terms of A, which gives you the following:

$$P(x) = |C|^2 e^{-2k_2x} = \frac{4k_1^2|A|^2 e^{-2k_2x}}{k_1^2 + k_2^2}$$

Note that although this value rapidly falls to zero as x gets large, it has a nonzero value near $x = 0$. This is an example of *quantum tunneling,* in which particles that wouldn't classically be allowed in certain regions can get there with quantum physics.

13 Calculate the transmission coefficient, T, for the wave function where $E > V_0$ and

$$V(x) = \begin{cases} 0 & x < 0 \\ V_0 & 0 \le x \le a \\ 0 & x > a \end{cases}$$

and where

- $\psi_1(x) = Ae^{ik_1x} + Be^{-ik_1x} \quad x < 0$
- $\psi_2(x) = Ce^{ik_2x} + De^{-ik_2x} \quad 0 \le x \le a$
- $\psi_3(x) = Ee^{ik_1x} \quad x > a$

where $k_1^2 = \frac{2mE}{\hbar^2}$ and $k_2^2 = \frac{2m(E-V_0)}{\hbar^2}$. **The answer is**

$$T = \left[1 + \frac{1}{4}\left(\frac{k_1^2 - k_2^2}{k_1 k_2}\right)^2 \sin^2(k_2 a)\right]^{-1}$$

where $k_1^2 = 2mE/\hbar^2$ and $k_2^2 = 2m(E - V_0)/\hbar^2$.

You know that

- $\psi_1(x) = Ae^{ik_1x} + Be^{-ik_1x} \quad x < 0$
- $\psi_2(x) = Ce^{k_2x} + De^{-k_2x} \quad 0 \le x \le a$
- $\psi_3(x) = Ee^{ik_1x} \quad x > a$

To determine A, B, C, D, and E, you use the boundary conditions, which work out here to be the following:

- $\psi_1(0) = \psi_2(0)$
- $\frac{d\psi_1}{dx}(0) = \frac{d\psi_2}{dx}(0)$
- $\psi_2(a) = \psi_3(a)$
- $\frac{d\psi_2}{dx}(a) = \frac{d\psi_3}{dx}(a)$

From the boundary conditions, you get the following by matching the values of wave functions and their derivatives:

- $A + B = C + D$
- $ik_1(A - B) = ik_2(C - D)$
- $Ce^{ik_2a} + De^{-ik_2a} = Ee^{ik_1a}$
- $ik_2Ce^{ik_2a} - ik_2De^{-ik_2a} = ik_1Ee^{ik_1a}$

So using these boundary conditions to solve for E gives you

$$E = 4k_1k_2Ae^{-ik_1a}\left[4k_1k_2\cos(k_2a) - 2i(k_1^2 + k_2^2)\sin(k_2a)\right]^{-1}$$

The transmission coefficient T is

$$T = \frac{|E|^2}{|A|^2}$$

Doing the algebra on the two previous equations gives you the answer:

$$T = \left[1 + \frac{1}{4}\left(\frac{k_1^2 - k_2^2}{k_1 k_2}\right)^2 \sin^2(k_2 a)\right]^{-1}$$

14 Calculate the reflection coefficient, R, for the wave function where $E > V_0$ and

$$V(x) = \begin{cases} 0 & x < 0 \\ V_0 & 0 \le x \le a \\ 0 & x > a \end{cases}$$

and where

- $\psi_1(x) = Ae^{ik_1x} + Be^{-ik_1x} \quad x < 0$
- $\psi_2(x) = Ce^{ik_2x} + De^{-ik_2x} \quad 0 \le x \le a$
- $\psi_3(x) = Ee^{ik_1x} \quad x > a$

where $k_1^2 = \frac{2mE}{\hbar^2}$ and $k_2^2 = \frac{2m(E - V_0)}{\hbar^2}$. **The answer is**

$$\mathbf{R} = \left[1 + \frac{4\left(\frac{E}{V_0}\right)\left(\frac{E}{V_0} - 1\right)}{\sin^2\left(a\left(2m\frac{V_0}{\hbar^2}\right)^{1/2}\left(\frac{E}{V_0} - 1\right)^{1/2}\right)}\right]^{-1}$$

You start from the wave equation

- $\psi_1(x) = Ae^{ik_1x} + Be^{-ik_1x} \quad x < 0$
- $\psi_2(x) = Ce^{ik_2x} + De^{-ik_2x} \quad 0 \le x \le a$
- $\psi_3(x) = Ee^{ik_1x} \quad x > a$

where $k_1^2 = 2m$ and $k_2^2 = 2m(E - V_0)/\hbar^2$. You can determine A, B, C, D, and E from the boundary conditions:

- $\psi_1(0) = \psi_2(0)$
- $\frac{d\psi_1}{dx}(0) = \frac{d\psi_2}{dx}(0)$
- $\psi_2(a) = \psi_3(a)$
- $\frac{d\psi_2}{dx}(a) = \frac{d\psi_3}{dx}(a)$

The reflection coefficient R is

$$R = \frac{|B|^2}{|A|^2}$$

Doing the math (using boundary conditions to solve for A and B) gives you

$$R = \left[1 + \frac{4\left(\frac{E}{V_0}\right)\left(\frac{E}{V_0} - 1\right)}{\sin^2\left(a\left(2m\frac{V_0}{\hbar^2}\right)^{1/2}\left(\frac{E}{V_0} - 1\right)^{1/2}\right)} \right]^{-1}$$

15 Find the wave function (up to arbitrary normalization coefficients) for a particle encountering the following potential barrier where the energy of the particle is $E < V_0$ (that is, the particle has less energy than the potential barrier this time)

$$V(x) = \begin{cases} 0 & x < 0 \\ V_0 & 0 \le x \le a \\ 0 & x > a \end{cases}$$

Here's the answer:

- $\psi_1(x) = Ae^{ik_1x} + Be^{-ik_1x} \quad x < 0$
- $\psi_2(x) = Ce^{k_2x} + De^{-k_2x} \quad 0 \le x \le a$
- $\psi_3(x) = Ee^{ik_1x} \quad x > a$

where $k_1^2 = 2mE/\hbar^2$ and $k_2^2 = 2m(E - V_0)/\hbar^2$.

In this case, the particle doesn't have as much energy as the potential of the barrier. Therefore, the Schrödinger equation looks like this for $x < 0$:

$$\psi_1(x) = Ae^{ik_1x} + Be^{-ik_1x} \quad x < 0$$

Here's what the Schrödinger equation looks like for $0 \le x \le a$:

$$\frac{d^2\psi_2}{dx^2}(x) + k_2^2\psi_2(x) = 0 \quad x > 0$$

where $k_2^2 = 2m(E - V_0)/\hbar^2$. Here, however, $E - V_0$ is less than 0, which would make k imaginary, which is impossible physically. So change the sign in the Schrödinger equation from plus to minus:

$$\frac{d^2\psi_2}{dx^2}(x) - k_2^2\psi_2(x) = 0 \quad x > 0$$

And use the following value for k_2:

$$k_2^2 = \frac{2m(V_0 - E)}{\hbar^2}$$

For the region $x > a$, here's what the Schrödinger equation looks like:

$$\frac{d^2\psi_3}{dx^2}(x) + k_1^2\psi_3(x) = 0 \quad x > a$$

where $k_1^2 = \frac{2mE}{\hbar^2}$.

Solve the differential equations to find the solutions for $\psi_1(x)$, $\psi_2(x)$, and $\psi_3(x)$, which are

- $\psi_1(x) = Ae^{ik_1x} + Be^{-ik_1x} \quad x < 0$
- $\psi_2(x) = Ce^{k_2x} + De^{-k_2x} \quad 0 \le x \le a$
- $\psi_3(x) = Ee^{ik_1x} + Fe^{-ik_1x} \quad x > a$

There's no leftward traveling wave in the region $x > a$; F = 0, so $\psi_3(x)$ is

$$\psi_3(x) = Ee^{ik_1x} \quad x > a$$

16 Calculate the reflection and transmission coefficients for a particle hitting a potential barrier of V_0 where the particle's energy $E < V_0$ and

$$V(x) = \begin{cases} 0 & x < 0 \\ V_0 & 0 \le x \le a \\ 0 & x > a \end{cases}$$

and where

- $\psi_1(x) = Ae^{ik_1x} + Be^{-ik_1x} \quad x < 0$
- $\psi_2(x) = Ce^{ik_2x} + De^{-ik_2x} \quad 0 \le x \le a$
- $\psi_3(x) = Ee^{ik_1x} \quad x > a$

where $k_1^2 = \frac{2mE}{\hbar^2}$ and $k_2^2 = \frac{2m(E - V_0)}{\hbar^2}$. **The answers are**

- $\mathbf{R} = \left(\frac{k_1^2 + k_2^2}{k_1 k_2}\right)^2 \sinh^2(k_2 a)\left[4\cosh^2(k_2 a) + \left(\frac{k_1^2 + k_2^2}{k_1 k_2}\right)^2 \sinh^2(k_2 a)\right]^{-1}$
- $\mathbf{T} = 4\left[4\cosh^2(k_2 a) + \left(\frac{k_1^2 - k_2^2}{k_1 k_2}\right)^2 \sinh^2(k_2 a)\right]^{-1}$

where $k_1^2 = 2mE/\hbar^2$ and $k_2^2 = 2m(V_0 - E)/\hbar^2$.

Start with this wave equation:

- $\psi_1(x) = Ae^{ik_1x} + Be^{-ik_1x} \quad x < 0$
- $\psi_2(x) = Ce^{k_2x} + De^{-k_2x} \quad 0 \le x \le a$
- $\psi_3(x) = Ee^{ik_1x} \quad x > a$

where $k_1^2 = 2mE/\hbar^2$ and $k_2^2 = 2m(V_0 - E)/\hbar^2$. The reflection and transmission coefficients are

- $R = \frac{|B|^2}{|A|^2}$
- $T = \frac{|E|^2}{|A|^2}$

You determine A, B, and E using these boundary conditions:

- $\psi_1(0) = \psi_2(0)$
- $\frac{d\psi_1}{dx}(0) = \frac{d\psi_2}{dx}(0)$
- $\psi_2(a) = \psi_3(a)$
- $\frac{d\psi_2}{dx}(a) = \frac{d\psi_3}{dx}(a)$

Doing the algebra to solve for R and T, and solving for A, B, and C using boundary conditions, gives you

- $R = \left(\frac{k_1^2 + k_2^2}{k_1 k_2}\right)^2 \sinh^2(k_2 a) \left[4\cosh^2(k_2 a) + \left(\frac{k_1^2 + k_2^2}{k_1 k_2}\right)^2 \sinh^2(k_2 a)\right]^{-1}$
- $T = 4\left[4\cosh^2(k_2 a) + \left(\frac{k_1^2 - k_2^2}{k_1 k_2}\right)^2 \sinh^2(k_2 a)\right]^{-1}$

Chapter 3

Over and Over with Harmonic Oscillators

In This Chapter

- Eyeing the total energy with the Hamiltonian operator
- Using raising and lowering operators to solve for energy states
- Examining how harmonic oscillator operators work with matrices

Harmonic oscillators are big in many branches of physics, including quantum physics. Classically, *harmonic oscillators* involve springs, pendulums, and other devices that you can measure with a stopwatch. When you're working with harmonic oscillators like springs and pendulums, you're looking at the macroscopic, classical picture. Springs and pendulums measured classically can have continuous energy levels, depending on the amount you stretch the spring or the angle at which you start the pendulum.

When you get to quantum physics, however, the story is different. The realm of quantum physics is the microscopic realm because that's where quantum physics effects become evident. In the case of harmonic oscillators, you deal with microscopic systems where a restoring force tends to keep particles in their position; for example, the atoms in a solid can act like harmonic oscillators, vibrating in place furiously, depending on the temperature of the solid. Each atom has kinetic energy, so it's in motion, but at the same time, electrostatic forces restore it to its original position, so it vibrates.

As always in quantum physics, you can't define the position or energy of harmonic oscillators on the quantum physics level with complete certainty, but you can describe those quantities in terms of probabilities — and that means you can work with probability amplitudes, as represented by wave functions. The energy levels of quantum physical harmonic oscillators are *quantized* — that is, only certain energy levels are allowed. And when you have wave functions, you can also have operators to work on those wave functions.

This chapter focuses on all aspects related to harmonic oscillators — on the quantum level — and helps you solve many types of problems. Note that I keep everything relatively simple in this chapter by restricting harmonic oscillators to one-dimensional motion. You can find a discussion of 3-D motion in Part III.

Total Energy: Getting On with a Hamiltonian

The first step in solving for wave functions is usually to create a Hamiltonian operator, which tells you the allowed energy states. The wave functions are the eigenstates of the Hamiltonian, and the allowed energy levels are the eigenvalues of the Hamiltonian, like this (see Chapter 1 for info on eigenvalues and eigenstates):

$$H|\psi\rangle = E|\psi\rangle$$

You can use a Hamiltonian with harmonic oscillators to help you find a system's energy level — that's what Hamiltonians are all about; they're the equations that give you the energy of the system.

The following example looks at a problem in classical terms, and the follow-up questions extrapolate into the quantum physics realm.

Q. As the first step toward describing harmonic oscillators in quantum physical terms, find the motion of classical harmonic oscillators. Assume the object in harmonic oscillator motion has mass m and the restoring force is $F = -kx$, where x is the distance the object is from its equilibrium position (where there's no net force acting on the object) and k is the spring constant.

A. $x = A\sin\omega t + B\cos\omega t$**, where** $k/m = \omega^2$**.**

Hooke's law tells you that the force on an object in harmonic oscillation is

$$F = -kx$$

Because by Newton's Second Law, $F = ma$, where m is the mass of the particle in harmonic motion and a is its instantaneous acceleration, you can substitute ma for F in the oscillation equation. Rewrite the equation as $F = -kx = ma$, or equivalently

$$ma + kx = 0$$

In classical physics, the definition for acceleration is

$$a = \frac{d^2x}{dt^2}$$

So you can also substitute for a. Rewrite the oscillation equation as

$$m\frac{d^2x}{dt^2} + kx = 0$$

Divide by the mass to get

$$\frac{d^2x}{dt^2} + \frac{kx}{m} = 0$$

For a classical harmonic oscillator, you know that $k/m = \omega^2$, where ω is the angular frequency; therefore, taking $k/m = \omega^2$ gives you

$$\frac{d^2x}{dt^2} + \omega^2 x = 0$$

Finally, solve this differential equation for x:

$$x = A\sin\omega t + B\cos\omega t$$

The solution is an oscillating one.

1. Determine the quantum physical Hamiltonian operator for harmonic oscillators, giving the total energy of the harmonic oscillator (that's the sum of the kinetic and potential energies) by adapting the classical energy equation so that position and momentum, x and p, become the operators X and P that return the position and momentum when applied to a wave function.

Solve It

2. Use the Hamiltonian from the preceding problem to create an energy eigenvector and energy eigenvalue equation for quantum physical harmonic oscillators.

Solve It

Up and Down: Using Some Crafty Operators

When handling quantum physics harmonic oscillator problems, you usually use one of two operators to make dealing with harmonic oscillators much easier (because you don't have to know the explicit form of the state vectors):

- **Raising operator (creation operator):** This operator raises the energy level of an eigenstate by one level. For example, if the harmonic oscillator is in the second energy level, this operator raises it one level to the third energy level.
- **Lowering operator (annihilation operator):** Although one of this operator's names makes it sound like a sequel to a sci-fi movie, this operator reduces eigenstates by one level.

Basically, these two operators allow you to comprehend the entire energy spectrum just by eyeing the energy difference between eigenstates.

So to start solving these harmonic oscillator problems, you first have to introduce two new operators, p and q, which are dimensionless. These operators relate to the P (momentum) operator and the X (location) operator this way:

- $p = \frac{P}{(m\hbar\omega)^{1/2}}$
- $q = X\left(\frac{m\omega}{\hbar}\right)^{1/2}$

You use these two new operators, p and q, as the basis of the lowering operator, a, and the raising operator, $a^\dagger$:

- $a = \frac{1}{\sqrt{2}}(q + ip)$
- $a^\dagger = \frac{1}{\sqrt{2}}(q - ip)$

Check out the following example using the raising and lowering operators; then try a couple of problems.

Q. Write the harmonic oscillator Hamiltonian in terms of the raising and lowering operators.

A. $\mathbf{H = \hbar\omega\left(a^\dagger a + \frac{1}{2}\right)}$**, where** $\frac{k}{m} = \omega^2$**.**

The Hamiltonian looks like this:

$$H = \frac{P^2}{2m} + \frac{1}{2}m\omega^2 X^2$$

And the p and q operators are equal to

- $p = \frac{P}{(m\hbar\omega)^{1/2}}$
- $q = X\left(\frac{m\omega}{\hbar}\right)^{1/2}$

The Hamiltonian and the p and q operators both involve P and X, which means you can perform some substitutions. First rearrange the p and q equations by solving for P and X:

- $(m\hbar\omega)^{1/2} p = P$
- $\frac{q}{(m\omega/\hbar)^{1/2}} = X$

Substitute the values of P and X into the Hamiltonian. Then use the fact that $a^\dagger a = \left(\frac{q^2 + p^2}{2}\right)$, where $a = \frac{1}{\sqrt{2}}(q + ip)$ and $a^\dagger = \frac{1}{\sqrt{2}}(q - ip)$, to get the answer:

$$H = \hbar\omega\left(a^\dagger a + \frac{1}{2}\right)$$

3. The $a^{\dagger}a$ operator, which is called the *number* operator (N), returns the number of the energy state that the harmonic oscillator is in (that is, 2 for the second excited state, 3 for the third excited state, and so on). Write the Hamiltonian in terms of N and write an equation showing how to use N on the harmonic oscillator eigenstates.

Solve It

4. What are the first three energy levels of a harmonic oscillator in terms of ω?

Solve It

Finding the Energy after Using the Raising and Lowering Operators

As in any quantized system, only certain energy levels are allowed for quantum harmonic oscillators. Finding those energy levels is the goal of this section. To find the energy, you designate a harmonic oscillator energy eigenvector like this: $|n\rangle$. This is the eigenvector for the nth energy level. When you apply the lowering operator, a, you're supposed to get the $|n-1\rangle$ eigenvector (up to an arbitrary constant): $a|n\rangle$.

How can you verify that you're getting the eigenvector $|n-1\rangle$? One way is to apply the Hamiltonian like this to measure the new energy eigenvector's energy: $H\left(a|n\rangle\right)$. This should turn out to be

$$H\left(a|n\rangle\right) = \left(E_n - \hbar\omega\right)\left(a|n\rangle\right)$$

To verify an expression like $H\left(a|n\rangle\right)$, you have to move the H operator past the a operator, which means you have to start by finding the commutator of a and H. (The *commutator* of operators A and B is [A, B] = AB – BA; see Chapter 1 for details.) This is an important step in finding the allowed energy levels.

Now look over the example problem and try the practice problems.

Q. What is the commutator of the lowering operator and the Hamiltonian $[a, H]$? And what is the commutator of the raising operator and the Hamiltonian $\left[a^\dagger, H\right]$?

A. $\boldsymbol{[a, H] = \hbar\omega a}$ **and** $\boldsymbol{\left[a^\dagger, H\right] = -\hbar\omega a^\dagger}$.

Start with the commutator of a and $a^\dagger$:

$$\left[a, a^\dagger\right] = \frac{1}{2}\left[q + ip,\ q - ip\right]$$

Through the definition of commutators, this becomes

$$\left[a, a^\dagger\right] = \frac{1}{2}\left[q + ip,\ q - ip\right] = -i\left[q, p\right]$$

which gives you the following:

$$\left[a, a^\dagger\right] = 1$$

In the example in the preceding section, you find that

$$H = \hbar\omega\left(a^\dagger a + \frac{1}{2}\right)$$

Therefore, you get the following answers by plugging in the value of H and simplifying:

- $\left[a, H\right] = \hbar\omega a$
- $\left[a^\dagger, H\right] = -\hbar\omega a^\dagger$

5. Evaluate $H\left(a|n\rangle\right)$ in terms of E_n, the energy of $|n\rangle$ (that is, $H|n\rangle = E_n|n\rangle$), given that $\left[a, H\right] = \hbar\omega a$.

Solve It

6. Evaluate $H\left(a^\dagger|n\rangle\right)$ in terms of E_n, the energy of $|n\rangle$ (that is, $H|n\rangle = E_n|n\rangle$), given that $\left[a^\dagger, H\right] = -\hbar\omega a$.

Solve It

Using the Raising and Lowering Operators Directly on the Eigenvectors

Sometimes you may need to apply the two operators directly to the eigenvectors, particularly to find the corresponding raised or lowered eigenvectors. To do so, first apply the Hamiltonian like this:

- $H\left(a|n\rangle\right) = \left(E_n - \hbar\omega\right)\left(a|n\rangle\right)$
- $H\left(a^\dagger|n\rangle\right) = \left(E_n + \hbar\omega\right)\left(a^\dagger|n\rangle\right)$

Now what if you were to use the raising and lowering operators directly on the eigenvectors to find the corresponding raised or lowered eigenvectors? You'd expect something like this:

- $a|n\rangle = C|n-1\rangle$
- $a^\dagger|n\rangle = D|n+1\rangle$

where C and D are positive constants. But what do C and D equal?

The tool you have in solving this question is a constraint: $|n\rangle$, $|n-1\rangle$, and $|n+1\rangle$ must all be normalized. What is that normalization condition? The following example shows you how to solve this type of problem, and the practice problems have you find C and D.

Q. What is the normalization condition that constrains $|n\rangle$, $|n-1\rangle$, and $|n+1\rangle$?

A. $\boldsymbol{\langle n|n\rangle = 1, \langle n-1|n-1\rangle = 1}$**, and** $\boldsymbol{\langle n+1|n+1\rangle = 1}$**.**

All eigenvectors must be normalized such that $\langle m|m\rangle = 1$. Set the bras and kets equal to 1 to get the answer.

7. Find C, where $a|n\rangle = C|n-1\rangle$.

Solve It

8. Find D, where $a^\dagger|n\rangle = D|n+1\rangle$.

Solve It

Finding the Harmonic Oscillator Ground State Wave Function

Finding the wave function is important because if you've found the wave function, you've found out how the particle in harmonic oscillation moves. This section starts with the ground state wave function. So just what is $|0\rangle$? Can't you get a spatial eigenstate of this eigenvector? Something like $\psi_0(x)$, not just $|0\rangle$?

In other words, you want to find $\langle x|0\rangle = \psi_0(x)$. To do so, you need an equation of some kind, and you can find that when you realize that when you apply the lowering operator to $|0\rangle$, you end up with 0. So the equation you need use to find the ground state wave function is $\langle x|a|0\rangle = 0$.

To work with the raising and lowering operators in position space, you need the representations of a and $a^\dagger$ in position space. Check out the following example for more help; then try the practice problems.

Q. What are the representations of a and $a^\dagger$ in position space?

A. $a = \frac{1}{x_0\sqrt{2}}\left(X + x_0^2 \frac{d}{dx}\right)$ **and**

$a^\dagger = \frac{1}{x_0\sqrt{2}}\left(X - x_0^2 \frac{d}{dx}\right)$, **where X is the position operator and** $x_0 = \left(\frac{\hbar}{mw}\right)^{1/2}$.

The p operator is defined as

$$p = \frac{P}{(m\hbar\omega)^{1/2}}$$

Because $P = -i\hbar\frac{d}{dx}$, you can substitute for the P in the p operator equation. Here's what this change gives you:

$$p = \frac{-i\hbar}{(m\hbar\omega)^{1/2}}\frac{d}{dx}$$

You know that $x_0 = \left(\frac{\hbar}{m\omega}\right)^{1/2}$, so you can make another substitution:

$$p = \frac{-i\hbar}{(m\hbar\omega)^{1/2}}\frac{d}{dx} = -ix_0\frac{d}{dx}$$

Now find the a operator. You know that $a = \frac{1}{\sqrt{2}}(q + ip)$ and that $q = X\left(\frac{m\omega}{\hbar}\right)^{1/2} = \frac{X}{x_0}$. Plug the value of q into the a operator. Here's what you get:

$$a = \frac{1}{\sqrt{2}}\left(\frac{X}{x_0} + x_0\frac{d}{dx}\right)$$

Finally, factor out $\frac{1}{x_0}$ and rewrite the equation as

$$a = \frac{1}{x_0\sqrt{2}}\left(X + x_0^2\frac{d}{dx}\right)$$

Similarly, $a^\dagger$ turns out to be this:

$$a^\dagger = \frac{1}{x_0\sqrt{2}}\left(X - x_0^2\frac{d}{dx}\right)$$

9. Find the ground state wave function — up to an arbitrary normalization constant — of the harmonic oscillator, $\psi_0(x)$, starting from $\langle x|a|0\rangle = 0$ (that is, the lowering operator on $|0\rangle$ must give 0).

Solve It

10. Normalize the ground state wave function $\psi_0(x)$. (For more on normalizing, see Chapter 2.)

Solve It

Finding the Excited States' Wave Functions

Sometimes you have to find the ground state wave function for a harmonic oscillator in higher states, $\psi_1(x)$, $\psi_2(x)$, and so on. To do so, just use the following well-known formula for the wave functions of a harmonic oscillator:

$$\psi_n(x)=\frac{1}{\pi^{1/4}\left(2^n n!x_0\right)^{1/2}}\ \mathrm{H}_n\left(\frac{x}{x_0}\right)\exp\left(\frac{-x^2}{2x_0^2}\right)$$

where $x_0=\left(\frac{\hbar}{m\omega}\right)^{1/2}$. But what's $\mathrm{H}_n(x)$? That's the nth *Hermite polynomial*, which is defined this way:

$$\mathrm{H}_n(x)=(-1)^n\ \exp\left(x^2\right)\frac{d^n}{dx^n}\ \exp\left(-x^2\right)$$

In this section, the example problem and practice problems deal with wave functions for excited states.

Note that when solving problems related to proton displacement, you may need to convert all length measurements into femtometers (1 fm = 1×10^{-15} m).

Q. Derive $\psi_1(x)$ without using the preceding formula for $\psi_n(x)$.

A. $\psi_1(x)=\frac{\sqrt{2}}{\pi^{1/4}x_0^{3/2}}\ x\exp\left(\frac{-x^2}{2x_0^2}\right)$, **where** $x_0=\left(\frac{\hbar}{m\omega}\right)^{1/2}$.

You know that $\psi_1(x)=\langle x|1\rangle$ and $|1\rangle=a^\dagger|0\rangle$. So replace $|1\rangle$ in the first equation, which gives you the following:

$$\psi_1(x)=\langle x|a^\dagger|0\rangle$$

Also, you know that $a^\dagger$ is

$$a^\dagger=\frac{1}{x_0\sqrt{2}}\left(\mathrm{X}-x_0^2\frac{d}{dx}\right)$$

Substitute for $a^\dagger$ in the wave function equation to get

$$\langle x|a^\dagger|0\rangle=\frac{1}{x_0\sqrt{2}}\langle x|\left(\mathrm{X}-x_0^2\frac{d}{dx}\right)|0\rangle$$

$$=\frac{1}{x_0\sqrt{2}}\left(\mathrm{X}-x_0^2\frac{d}{dx}\right)\langle x|0\rangle$$

Because $\langle x|0\rangle=\psi_0(x)$, replace $\langle x|0\rangle$ in the preceding equation:

$$\psi_1(x)=\frac{1}{x_0\sqrt{2}}\left(\mathrm{X}-x_0^2\frac{d}{dx}\right)\psi_0(x)$$

Now find the derivative and simplify:

$$\psi_1(x)=\frac{1}{x_0\sqrt{2}}\left(x-x_0^2\frac{(-x)}{x_0^2}\right)\psi_0(x)$$

which becomes

$$\psi_1(x)=\frac{\sqrt{2}}{x_0}x\psi_0(x)$$

You know that

$$\psi_0(x)=\frac{\sqrt{2}}{\pi^{1/4}x_0^{1/2}}\ \exp\left(\frac{-x^2}{2x_0^2}\right)$$

Plug in the value for $\psi_0(x)$ into the equation for the first excited state, which gives you the answer:

$$\psi_1(x)=\frac{\sqrt{2}}{\pi^{1/4}x_0^{3/2}}\ x\exp\left(\frac{-x^2}{2x_0^2}\right)$$

11. What is the differential equation that describes $\psi_2(x)$ in terms of $\psi_0(x)$?

Solve It

12. What are the first six Hermite polynomials?

Solve It

13. Say that you have a proton undergoing harmonic oscillation with $\omega = 4.58 \times 10^{21}$ sec^{-1}. What are the first four energy levels?

Solve It

14. Say that you have a proton undergoing harmonic oscillation with $\omega = 4.58 \times 10^{21}$ sec^{-1}. What are the first three wave functions?

Solve It

Looking at Harmonic Oscillators in Matrix Terms

Harmonic oscillators have regularly spaced energy levels, so they're often viewed at in terms of matrices. For example, the following matrix may be the ground state eigenvector $|0\rangle$ (for an infinite vector):

$$|0\rangle = \begin{vmatrix} 1 \\ 0 \\ 0 \\ 0 \\ 0 \\ 0 \\ . \\ . \\ . \end{vmatrix}$$

And this may be the second excited state, $|2\rangle$:

$$|2\rangle = \begin{vmatrix} 0 \\ 0 \\ 1 \\ 0 \\ 0 \\ 0 \\ . \\ . \\ . \end{vmatrix}$$

You can handle harmonic oscillators using matrices instead of wave functions. The following example shows you how to do so.

Q. What is the number operator, N, in matrix terms? Demonstrate that N works properly on the $|2\rangle$ eigenvector.

A.

$$\mathbf{N} = \begin{vmatrix} \mathbf{0} & \mathbf{0} & \mathbf{0} & \mathbf{0} & \mathbf{.} & \mathbf{.} & \mathbf{.} \\ \mathbf{0} & \mathbf{1} & \mathbf{0} & \mathbf{0} & \mathbf{.} & \mathbf{.} & \mathbf{.} \\ \mathbf{0} & \mathbf{0} & \mathbf{2} & \mathbf{0} & \mathbf{.} & \mathbf{.} & \mathbf{.} \\ \mathbf{0} & \mathbf{0} & \mathbf{0} & \mathbf{3} & \mathbf{.} & \mathbf{.} & \mathbf{.} \\ \mathbf{0} & \mathbf{0} & \mathbf{0} & \mathbf{0} & \mathbf{.} & \mathbf{.} & \mathbf{.} \\ \mathbf{0} & \mathbf{0} & \mathbf{0} & \mathbf{0} & \mathbf{.} & \mathbf{.} & \mathbf{.} \\ \mathbf{.} & \mathbf{.} & \mathbf{.} & \mathbf{.} & \mathbf{.} & \mathbf{.} & \mathbf{.} \\ \mathbf{.} & \mathbf{.} & \mathbf{.} & \mathbf{.} & \mathbf{.} & \mathbf{.} & \mathbf{.} \\ \mathbf{.} & \mathbf{.} & \mathbf{.} & \mathbf{.} & \mathbf{.} & \mathbf{.} & \mathbf{.} \end{vmatrix}$$

The N operator, which just returns the energy level, would look like this:

$$N = \begin{vmatrix} 0 & 0 & 0 & 0 & . & . & . \\ 0 & 1 & 0 & 0 & . & . & . \\ 0 & 0 & 2 & 0 & . & . & . \\ 0 & 0 & 0 & 3 & . & . & . \\ 0 & 0 & 0 & 0 & . & . & . \\ 0 & 0 & 0 & 0 & . & . & . \\ . & . & . & . & . & . & . \\ . & . & . & . & . & . & . \\ . & . & . & . & . & . & . \end{vmatrix}$$

Here's what $N|2\rangle$ gives you:

$$N|2\rangle = \begin{vmatrix} 0 & 0 & 0 & 0 & . & . & . \\ 0 & 1 & 0 & 0 & . & . & . \\ 0 & 0 & 2 & 0 & . & . & . \\ 0 & 0 & 0 & 3 & . & . & . \\ 0 & 0 & 0 & 0 & . & . & . \\ 0 & 0 & 0 & 0 & . & . & . \\ . & . & . & . & . & . & . \\ . & . & . & . & . & . & . \\ . & . & . & . & . & . & . \end{vmatrix} \begin{vmatrix} 0 \\ 0 \\ 1 \\ 0 \\ 0 \\ 0 \\ . \\ . \\ . \end{vmatrix}$$

Doing the matrix multiplication gives you the following:

$$N|2\rangle = \begin{vmatrix} 0 \\ 0 \\ 2 \\ 0 \\ 0 \\ 0 \\ . \\ . \\ . \end{vmatrix} = \begin{vmatrix} 0 & 0 & 0 & 0 & . & . & . \\ 0 & 1 & 0 & 0 & . & . & . \\ 0 & 0 & 2 & 0 & . & . & . \\ 0 & 0 & 0 & 3 & . & . & . \\ 0 & 0 & 0 & 0 & . & . & . \\ 0 & 0 & 0 & 0 & . & . & . \\ . & . & . & . & . & . & . \\ . & . & . & . & . & . & . \\ . & . & . & . & . & . & . \end{vmatrix} \begin{vmatrix} 0 \\ 0 \\ 1 \\ 0 \\ 0 \\ 0 \\ . \\ . \\ . \end{vmatrix}$$

In other words, $N|2\rangle = 2|2\rangle$, which is what you'd expect.

15. What is the lowering operator, a, in matrix terms? Verify that it worked on the $|1\rangle$ state.

Solve It

16. What is the raising operator, $a^\dagger$, in matrix terms? Verify that it worked on the $|1\rangle$ state.

Solve It

Answers to Problems on Harmonic Oscillators

The following are the answers to the practice questions presented in this chapter. You can see the original questions, and answers in bold, and the answers worked out, step by step.

1 Determine the quantum physical Hamiltonian operator for harmonic oscillators, giving the total energy of the harmonic oscillator (that's the sum of the kinetic and potential energies) by adapting the classical energy equation so that position and momentum, *x* and *p,* become the operators X and P that return the position and momentum when applied to a wave function. **Here's the answer:**

$$\mathbf{H}=\frac{\mathbf{P}^2}{2m}+\frac{1}{2}m\omega^2\mathbf{X}^2$$

The Hamiltonian is the sum of the kinetic and potential energies:

$$H = KE + PE$$

To solve this problem, you need to know the kinetic and potential energy in terms of momentum and position. The kinetic energy at any one moment is

$$KE=\frac{p^2}{2m}$$

where *p* is the particle's momentum. And the particle's potential energy is equal to

$$PE=\frac{1}{2}kx^2=\frac{1}{2}m\omega^2x^2$$

where *k* is the spring constant and $k/m=\omega^2$. Now replace *p* and *x* with the P and X operators in the Hamiltonian, KE + PE:

$$H=\frac{P^2}{2m}+\frac{1}{2}m\omega^2X^2$$

2 Use the Hamiltonian from the preceding problem to create an energy eigenvector and energy eigenvalue equation for quantum physical harmonic oscillators.

$$\mathbf{H}|\psi\rangle=\frac{\mathbf{P}^2}{2m}|\psi\rangle+\frac{1}{2}m\omega^2\mathbf{X}^2|\psi\rangle=\mathbf{E}|\psi\rangle$$

You use the Hamiltonian with eigenvectors and eigenvalues like this:

$$H|\psi\rangle=E|\psi\rangle$$

where E is the energy eigenvalue of the eigenstate $|\psi\rangle$. Now rewrite $H|\psi\rangle$ using the Hamiltonian equation from problem 1 to get the following:

$$H|\psi\rangle=\frac{P^2}{2m}|\psi\rangle+\frac{1}{2}m\omega^2X^2|\psi\rangle=E|\psi\rangle$$

3 The $a^\dagger a$ operator, which is called the *number* operator (N), returns the number of the energy state that the harmonic oscillator is in (that is, 2 for the second excited state, 3 for the third excited state, and so on). Write the Hamiltonian in terms of N and write an equation showing how to use N on the harmonic oscillator eigenstates. **The answers are**

$$\mathbf{H} = \hbar\omega\left(\mathbf{N}+\frac{1}{2}\right) \text{ and } \mathbf{N}|n\rangle = n|n\rangle$$

where $|n\rangle$ is the *n*th excited state.

Given that $H = \hbar\omega\left(a^\dagger a + \frac{1}{2}\right)$ and that $a^\dagger a = N$, you can write the Hamiltonian H as

$$H = \hbar\omega\left(N+\frac{1}{2}\right)$$

Now denote the eigenstates of N as $|n\rangle$, which gives you this:

$$N|n\rangle = n|n\rangle$$

where n is the number of the nth state.

4 What are the first three energy levels of a harmonic oscillator in terms of ω? **Here are the answers:**

- $\mathbf{E_0} = \frac{1}{2}\hbar\omega$
- $\mathbf{E_1} = \frac{3}{2}\hbar\omega$
- $\mathbf{E_2} = \frac{5}{2}\hbar\omega$

Because $H = \hbar\omega\left(N+\frac{1}{2}\right)$ and $H|n\rangle = E_n|n\rangle$, you can set E_n equal to the value of H:

$$E_n = \left(n+\frac{1}{2}\right)\hbar\omega \quad n = 0, 1, 2, \ldots$$

Now just plug in the numbers for n and simplify to find the first three energy levels. The energy of the ground state ($n = 0$) is

$$E_0 = \frac{1}{2}\hbar\omega$$

And the first excited state ($n = 1$) is

$$E_1 = \frac{3}{2}\hbar\omega$$

And the second excited state ($n = 2$) has an energy of

$$E_2 = \frac{5}{2}\hbar\omega$$

5 Evaluate $H(a|n\rangle)$ in terms of E_n, the energy of $|n\rangle$ (that is, $H|n\rangle = E_n|n\rangle$), given that $[a, H] = \hbar\omega a$.

$\mathbf{H(a|n\rangle) = (E_n - \hbar\omega)(a|n\rangle)}$.

You want to find

$$H(a|n\rangle)$$

Using the commutator of $[a, \mathrm{H}]$ to move the H operator closer to $|n\rangle$, this equals

$$\mathrm{H}\left(a|n\rangle\right) = \left(a\mathrm{H} - \hbar\omega a\right)|n\rangle$$

So applying the Hamiltonian operator H, you get

$$\mathrm{H}\left(a|n\rangle\right) = \left(a\mathrm{H} - \hbar\omega a\right)|n\rangle = \left(\mathrm{E}_n - \hbar\omega\right)\left(a|n\rangle\right)$$

Therefore, $a|n\rangle$ is also an eigenvector of the harmonic oscillator, with energy $\mathrm{E}_n - \hbar\omega$.

6 Evaluate $\mathrm{H}\left(a^\dagger|n\rangle\right)$ in terms of E_n, the energy of $|n\rangle$ (that is, $H|n\rangle = \mathrm{E}_n|n\rangle$), given that $\left[a^\dagger, \mathrm{H}\right] = -\hbar\omega a$. $\mathbf{H}\left(\boldsymbol{a}^\dagger|\boldsymbol{n}\rangle\right) = \left(\boldsymbol{E}_n + \boldsymbol{\hbar\omega}\right)\left(\boldsymbol{a}^\dagger|\boldsymbol{n}\rangle\right)$.

You want to find

$$\mathrm{H}\left(a^\dagger|n\rangle\right)$$

Using the commutator of $\left[a^\dagger, \mathrm{H}\right]$ to get H to work on $|n\rangle$, this equals

$$\mathrm{H}\left(a^+|n\rangle\right) = \left(a^+\mathrm{H} - \hbar\omega a^+\right)|n\rangle$$

Apply the Hamiltonian:

$$\mathrm{H}\left(a^\dagger|n\rangle\right) = \left(a^\dagger\mathrm{H} + \hbar\omega a^\dagger\right)|n\rangle = \left(\mathrm{E}_n + \hbar\omega\right)\left(a^\dagger|n\rangle\right)$$

This means that $a^\dagger|n\rangle$ is an eigenstate of the harmonic oscillator, with energy $\mathrm{E}_n + \hbar\omega$, not just E_n.

7 Find C, where $a|n\rangle = \mathrm{C}|n-1\rangle$. $\mathbf{C} = \boldsymbol{n}^{1/2}$.

Start with this expression:

$$\left(\langle n|a^\dagger\right)\left(a|n\rangle\right)$$

This expression equals the following, by the definition of the raising and lowering operators:

$$\left(\langle n|a^\dagger\right)\left(a|n\rangle\right) = \mathrm{C}^2\langle n-1|n-1\rangle$$

And because $|n-1\rangle$ is normalized, $\langle n-1|n-1\rangle = 1$. Simplify the right side of the equation:

$$\left(\langle n|a^\dagger\right)\left(a|n\rangle\right) = \mathrm{C}^2$$

You also know that $a^\dagger a = \mathrm{N}$, the energy level operator, and that

$$\left(\langle n|a^\dagger\right)\left(a|n\rangle\right) = \langle n|a^\dagger a|n\rangle$$

Because $a^\dagger a = \mathrm{N}$, you know that

$$\left(\langle n|a^\dagger\right)\left(a|n\rangle\right) = \langle n|a^\dagger a|n\rangle = \langle n|\mathrm{N}|n\rangle$$

Because $\mathrm{N}|n\rangle = n|n\rangle$, where n is the energy level, you get the following:

$$\left(\langle n|a^\dagger\right)\left(a|n\rangle\right) = \langle n|\mathrm{N}|n\rangle = n\langle n|n\rangle$$

But $\langle n|n\rangle = 1$, so

$$\left(\langle n|a^\dagger\right)\left(a|n\rangle\right) = n\langle n|n\rangle = n$$

Because you know that $\left(\langle n|a^\dagger\right)\left(a|n\rangle\right) = C^2$ and also that $\left(\langle n|a^\dagger\right)\left(a|n\rangle\right) = n$, you get

$$C^2 = n, \text{ or } C = n^{1/2}$$

Therefore, $a|n\rangle = n^{1/2}|n-1\rangle$.

8 Find D, where $a^\dagger|n\rangle = D|n+1\rangle$. **D = $(n + 1)^{1/2}$.**

Start with this expression:

$$\left(\langle n|a\right)\left(a^\dagger|n\rangle\right)$$

which equals

$$\left(\langle n|a\right)\left(a^\dagger|n\rangle\right) = D^2\langle n+1|n+1\rangle$$

And because $|n+1\rangle$ is normalized, $\langle n+1|n+1\rangle = 1$, so

$$\left(\langle n|a\right)\left(a^\dagger|n\rangle\right) = D^2$$

You know that $a^\dagger a = N$, the energy level operator, and also that $\left(\langle n|a\right)\left(a^\dagger|n\rangle\right) = \langle n|aa^\dagger|n\rangle$. Use the fact that $\left[a, a^\dagger\right] = 1$ to get

$$\left(\langle n|a\right)\left(a^\dagger|n\rangle\right) = \langle n|aa^\dagger|n\rangle = \langle n|a^\dagger a + 1|n\rangle$$

Because $a^\dagger a = N$, you know that

$$\left(\langle n|a\right)\left(a^\dagger|n\rangle\right) = \langle n|a^\dagger a + 1|n\rangle = \langle n|N + 1|n\rangle$$

Because $N|n\rangle = n|n\rangle$, where n is the energy level, you get the following:

$$\left(\langle n|a\right)\left(a^\dagger|n\rangle\right) = \langle n|N + 1|n\rangle = (n+1)\langle n|n\rangle$$

But $\langle n|n\rangle = 1$, so

$$\left(\langle n|a\right)\left(a^\dagger|n\rangle\right) = (n+1)\langle n|n\rangle = n+1$$

Because $\left(\langle n|a\right)\left(a^\dagger|n\rangle\right) = D^2$ and $\left(\langle n|a\right)\left(a^\dagger|n\rangle\right) = n+1$, you get

$$D^2 = n + 1$$

$$D = (n + 1)^{1/2}$$

which means

$$a^\dagger|n\rangle = (n+1)^{1/2}|n+1\rangle$$

9 Find the ground state wave function — up to an arbitrary normalization constant — of the harmonic oscillator, $\psi_0(x)$, starting from $\langle x|a|0\rangle = 0$ (that is, the lowering operator on $|0\rangle$ must give 0). $\boldsymbol{\psi_0(x) = A\exp\left(\frac{-x^2}{2x_0^2}\right)}$.

Start with this equation:

$$a|0\rangle = 0$$

Apply the $\langle x|$ bra, which gives you

$$\langle x|a|0\rangle = 0$$

Substitute for a using its position representation:

$$\langle x|a|0\rangle = \frac{1}{x_0\sqrt{2}}\langle x\left|X + x_0^2\frac{d}{dx}\right|0\rangle = 0$$

You know that $\langle x|0\rangle = \psi_0(x)$, so replace $\langle x|$ and $|0\rangle$ with $\psi_0(x)$. Distributing gives you the following:

$$\langle x|a|0\rangle = \frac{1}{x_0\sqrt{2}}\left(x\psi_0(x) + x_0^2\frac{d\psi_0(x)}{dx}\right) = 0$$

Now multiply both sides by $x_0\sqrt{2}$, which gives you

$$x\psi_0 + x_0^2\frac{d\psi_0(x)}{dx} = 0$$

Isolate the differential equation on one side of the equal sign:

$$\frac{d\psi_0(x)}{dx} = -\frac{x}{x_0^2}\psi_0(x)$$

Finally, solve this differential equation. Here's what you get:

$$\psi_0(x) = \text{A}\exp\left(\frac{-x^2}{2x_0^2}\right)$$

10 Normalize the ground state wave function $\psi_0(x)$.

$\mathbf{A} = \frac{1}{\pi^{1/4}x_0^{1/2}}$, so

$$\boldsymbol{\psi_0(x) = \frac{1}{\pi^{1/4}x_0^{1/2}}\exp\left(\frac{-x^2}{2x_0^2}\right)}$$

If $\psi_0(x) = \text{A}\exp\left(\frac{-x^2}{2x_0^2}\right)$, you want to find A. Wave functions must be normalized, so you know that the following equation must be true:

$$1 = \int_{-\infty}^{\infty}|\psi_0(x)|^2\,dx$$

Substituting for $\psi_0(x)$ gives you

$$1 = \int_{-\infty}^{\infty}\left|\text{A}\exp\left(\frac{-x^2}{2x_0^2}\right)\right|^2 dx$$

Move the constant A outside of the integral:

$$1 = \text{A}^2\int_{-\infty}^{\infty}\exp\left(\frac{-x^2}{x_0^2}\right)dx$$

Evaluate the integral to get the following:

$$1 = A^2 \int_{-\infty}^{\infty} \exp\left(\frac{-x^2}{x_0^2}\right) dx = A^2 \pi^{1/2} x_0$$

Take the equation

$$1 = A^2 \pi^{1/2} x_0$$

and solve for A:

$$A = \frac{1}{\pi^{1/4} x_0^{1/2}}$$

Now fill in the value of A in the wave function for the ground state of a harmonic oscillator. Here's what you get:

$$\psi_0(x) = \frac{1}{\pi^{1/4} x_0^{1/2}} \exp\left(\frac{-x^2}{2x_0^2}\right)$$

11 What is the differential equation that describes $\psi_2(x)$ in terms of $\psi_0(x)$?

$$\boldsymbol{\psi_2(x) = \frac{1}{\sqrt{2!}} \frac{1}{2!x_0^2} \left(x - x_0^2 \frac{d}{dx}\right)^2 \psi_0(x)}$$

You can find $\psi_2(x)$ from this equation:

$$\psi_2(x) = \frac{1}{\sqrt{2!}} \langle x | (a^\dagger)^2 | 0 \rangle$$

Substitute for $a^\dagger$, which gives you

$$\psi_2(x) = \frac{1}{\sqrt{2!}} \frac{1}{2!x_0^2} \left(x - x_0^2 \frac{d}{dx}\right)^2 \psi_0(x)$$

You can generalize this differential equation for $\psi_n(x)$ like this:

$$\psi_n(x) = \frac{1}{\pi^{1/4} (2^n n!)^{1/2}} \frac{1}{x_0^{n+1/2}} \left(x - x_0^2 \frac{d}{dx}\right)^n \exp\left(\frac{-x^2}{x_0^2}\right)$$

12 What are the first six hermite polynomials? **Here are your answers:**

- **$H_0(x) = 1$**
- **$H_1(x) = 2x$**
- **$H_2(x) = 4x^2 - 2$**
- **$H_3(x) = 8x^3 - 12x$**
- **$H_4(x) = 16x^4 - 48x^2 + 12$**
- **$H_5(x) = 32x^5 - 160x^3 + 120x$**

Here's how to find $H_n(x)$:

$$H_n(x) = (-1)^n \exp(x^2) \frac{d^n}{dx^n} \exp(-x^2)$$

To find the first six hermite polynomials, simply perform the calculations for $n = 0$ through $n = 5$.

13 Say that you have a proton undergoing harmonic oscillation with $\omega = 4.58 \times 10^{21}\ \text{sec}^{-1}$. What are the first four energy levels? **The answers are**

- $\mathbf{E_0 = \frac{\hbar\omega}{2} = 1.50\ MeV}$
- $\mathbf{E_1 = \frac{3\hbar\omega}{2} = 4.50\ MeV}$
- $\mathbf{E_2 = \frac{5\hbar\omega}{2} = 7.50\ MeV}$
- $\mathbf{E_3 = \frac{7\hbar\omega}{2} = 10.50\ MeV}$

You know that in general

$$E_n = \left(n + \frac{1}{2}\right)\hbar\omega \quad n = 0, 1, 2, \ldots$$

To find the first four energy levels of the proton, perform the calculations for $n = 0$ through $n = 3$.

14 Say that you have a proton undergoing harmonic oscillation with $\omega = 4.58 \times 10^{21}\ \text{sec}^{-1}$. What are the first three wave functions? **Here are the answers:**

- $\boldsymbol{\psi_0(x) = \frac{1}{1.92\pi^{1/4}} \exp\left(\frac{-x^2}{27.5}\right)}$
- $\boldsymbol{\psi_1(x) = \frac{1}{2.72\pi^{1/4}} 2\left(\frac{x}{3.71}\right) \exp\left(\frac{-x^2}{27.5}\right)}$
- $\boldsymbol{\psi_2(x) = \frac{1}{5.45\pi^{1/4}} \left(4\left(\frac{x}{3.71}\right)^2 - 2\right) \exp\left(\frac{-x^2}{27.5}\right)}$

where x is measured in femtometers (fm).

The general form of $\psi_n(x)$ is

$$\psi_n(x) = \frac{1}{\pi^{1/4}\left(2^n n! x_0\right)} H_n\left(\frac{x}{x_0}\right) \exp\left(\frac{-x^2}{2x_0^2}\right)$$

where $x_0 = \left(\hbar/(m\omega)\right)^{1/2}$. Plug in the numbers, $x_0 = 3.71 \times 10^{-15}$ m, and convert all length measurements into femtometers ($1\ \text{fm} = 1 \times 10^{-15}$ m), giving you $x_0 = 3.71$ fm. That makes $\psi_0(x)$ equal to

$$\psi_0(x) = \frac{1}{1.92\pi^{1/4}} \exp\left(\frac{-x^2}{27.5}\right)$$

where x is measured in fm.

Now plug in $n = 1$ and $n = 2$ into the general form $\psi_n(x)$ to find the first two excited states:

- $\psi_1(x) = \frac{1}{2.72\pi^{1/4}} \; 2\left(\frac{x}{3.71}\right)\exp\left(\frac{-x^2}{27.5}\right)$
- $\psi_2(x) = \frac{1}{5.45\pi^{1/4}} \left(4\left(\frac{x}{3.71}\right)^2 - 2\right)\exp\left(\frac{-x^2}{27.5}\right)$

15 What is the lowering operator, a, in matrix terms? Verify that it worked on the $|1\rangle$ state. **The answer is**

$$\boldsymbol{a} = \begin{vmatrix} \mathbf{0} & \sqrt{\mathbf{1}} & \mathbf{0} & \mathbf{0} & \cdots \\ \mathbf{0} & \mathbf{0} & \sqrt{\mathbf{2}} & \mathbf{0} & \cdots \\ \mathbf{0} & \mathbf{0} & \mathbf{0} & \sqrt{\mathbf{3}} & \cdots \\ \mathbf{0} & \mathbf{0} & \mathbf{0} & \mathbf{0} & \cdots \\ \mathbf{0} & \mathbf{0} & \mathbf{0} & \mathbf{0} & \cdots \\ \mathbf{0} & \mathbf{0} & \mathbf{0} & \mathbf{0} & \cdots \\ \cdot & \cdot & \cdot & \cdot & \cdots \\ \cdot & \cdot & \cdot & \cdot & \cdots \\ \cdot & \cdot & \cdot & \cdot & \cdots \end{vmatrix}$$

The lowering operator looks like this:

$$a = \begin{vmatrix} 0 & \sqrt{1} & 0 & 0 & \cdots \\ 0 & 0 & \sqrt{2} & 0 & \cdots \\ 0 & 0 & 0 & \sqrt{3} & \cdots \\ 0 & 0 & 0 & 0 & \cdots \\ 0 & 0 & 0 & 0 & \cdots \\ 0 & 0 & 0 & 0 & \cdots \\ \cdot & \cdot & \cdot & \cdot & \cdots \\ \cdot & \cdot & \cdot & \cdot & \cdots \\ \cdot & \cdot & \cdot & \cdot & \cdots \end{vmatrix}$$

The lowering operator should give you

$$a|1\rangle = n^{1/2}|n-1\rangle$$

And $a|1\rangle$ is

$$a|1\rangle \begin{vmatrix} 0 & \sqrt{1} & 0 & 0 & \cdots \\ 0 & 0 & \sqrt{2} & 0 & \cdots \\ 0 & 0 & 0 & \sqrt{3} & \cdots \\ 0 & 0 & 0 & 0 & \cdots \\ 0 & 0 & 0 & 0 & \cdots \\ 0 & 0 & 0 & 0 & \cdots \\ \cdot & \cdot & \cdot & \cdot & \cdots \\ \cdot & \cdot & \cdot & \cdot & \cdots \\ \cdot & \cdot & \cdot & \cdot & \cdots \end{vmatrix} \begin{vmatrix} 0 \\ 1 \\ 0 \\ 0 \\ 0 \\ 0 \\ \cdot \\ \cdot \\ \cdot \end{vmatrix}$$

Now perform the matrix multiplication. Here's what you get:

$$a|1\rangle = \begin{vmatrix} \sqrt{1} \\ 0 \\ 0 \\ 0 \\ 0 \\ 0 \\ \vdots \end{vmatrix} = \begin{vmatrix} 0 & \sqrt{1} & 0 & 0 & \cdots \\ 0 & 0 & \sqrt{2} & 0 & \cdots \\ 0 & 0 & 0 & \sqrt{3} & \cdots \\ 0 & 0 & 0 & 0 & \cdots \\ 0 & 0 & 0 & 0 & \cdots \\ 0 & 0 & 0 & 0 & \cdots \\ \vdots & \vdots & \vdots & \vdots & \ddots \end{vmatrix} \begin{vmatrix} 0 \\ 1 \\ 0 \\ 0 \\ 0 \\ 0 \\ \vdots \end{vmatrix}$$

In other words, $a|1\rangle = |0\rangle$, just as expected.

16 What is the raising operator, $a^\dagger$, in matrix terms? Verify that it worked on the $|1\rangle$ state. **The answer is**

$$\boldsymbol{a^\dagger} = \begin{vmatrix} \mathbf{0} & \mathbf{0} & \mathbf{0} & \mathbf{0} & \cdots \\ \sqrt{\mathbf{1}} & \mathbf{0} & \mathbf{0} & \mathbf{0} & \cdots \\ \mathbf{0} & \sqrt{\mathbf{2}} & \mathbf{0} & \mathbf{0} & \cdots \\ \mathbf{0} & \mathbf{0} & \sqrt{\mathbf{3}} & \mathbf{0} & \cdots \\ \mathbf{0} & \mathbf{0} & \mathbf{0} & \sqrt{\mathbf{4}} & \cdots \\ \mathbf{0} & \mathbf{0} & \mathbf{0} & \mathbf{0} & \cdots \\ \vdots & \vdots & \vdots & \vdots & \ddots \end{vmatrix}$$

Here's how the raising operator works in general:

$$a^\dagger|n\rangle = (n+1)^{1/2}|n+1\rangle$$

In matrix terms, $a^\dagger$ looks like this:

$$a^\dagger = \begin{vmatrix} 0 & 0 & 0 & 0 & \cdots \\ \sqrt{1} & 0 & 0 & 0 & \cdots \\ 0 & \sqrt{2} & 0 & 0 & \cdots \\ 0 & 0 & \sqrt{3} & 0 & \cdots \\ 0 & 0 & 0 & \sqrt{4} & \cdots \\ 0 & 0 & 0 & 0 & \cdots \\ \vdots & \vdots & \vdots & \vdots & \ddots \end{vmatrix}$$

You expect that $a^\dagger|1\rangle = \sqrt{2}|1\rangle$. The matrix multiplication is

$$a^\dagger|1\rangle = \begin{vmatrix} 0 & 0 & 0 & 0 & \cdots \\ \sqrt{1} & 0 & 0 & 0 & \cdots \\ 0 & \sqrt{2} & 0 & 0 & \cdots \\ 0 & 0 & \sqrt{3} & 0 & \cdots \\ 0 & 0 & 0 & \sqrt{4} & \cdots \\ 0 & 0 & 0 & 0 & \cdots \\ \vdots & \vdots & \vdots & \vdots & \vdots \end{vmatrix} \begin{vmatrix} 0 \\ 1 \\ 0 \\ 0 \\ 0 \\ 0 \\ \vdots \end{vmatrix}$$

And this equals

$$a^\dagger|1\rangle = \begin{vmatrix} 0 \\ 0 \\ \sqrt{2} \\ 0 \\ 0 \\ 0 \\ \vdots \end{vmatrix} = \begin{vmatrix} 0 & 0 & 0 & 0 & \cdots \\ \sqrt{1} & 0 & 0 & 0 & \cdots \\ 0 & \sqrt{2} & 0 & 0 & \cdots \\ 0 & 0 & \sqrt{3} & 0 & \cdots \\ 0 & 0 & 0 & \sqrt{4} & \cdots \\ 0 & 0 & 0 & 0 & \cdots \\ \vdots & \vdots & \vdots & \vdots & \vdots \end{vmatrix} \begin{vmatrix} 0 \\ 1 \\ 0 \\ 0 \\ 0 \\ 0 \\ \vdots \end{vmatrix}$$

So $a^\dagger\rangle = \sqrt{2}|2\rangle$, as it should.

Part II
Round and Round with Angular Momentum and Spin

"I really don't think showing formulas for angular momentum is going to enhance the spinning top experience for him."

In this part . . .

Quantum physics puts its own spin on the angular momentum problems you see in classical physics. In fact, particles have rotational energy in addition to kinetic and potential energy, and electrons have a kind of intrinsic angular momentum called *spin.* Angular momentum and spin are important topics in quantum physics because they're both quantized — a concept that classical physics can't handle. In this part, you get to tackle problems on each topic.

Chapter 4

Handling Angular Momentum in Quantum Physics

In This Chapter

- Working with angular momentum
- Using angular momentum in the Hamiltonian
- Finding the matrix representation of angular momentum
- Determining the eigenfunctions of angular momentum

In addition to linear kinetic and potential energy (see Chapters 2 and 3), quantum physics also deals with the energy associated with angular momentum if an item is spinning. When handling angular momentum, you start with the Hamiltonian, just as with linear energy types. In this case of angular momentum, the Hamiltonian looks like this:

$$H = \frac{L^2}{2I}$$

Here, L is the angular momentum operator and I is the rotational moment of inertia.

Angular momentum is a vector in three-dimensional space; it can be pointing any direction. Angular momentum is usually given by a magnitude and a component in one direction, which is usually the Z direction. So in addition to the magnitude l, you also specify the component of **L** in the Z direction, L_z. (***Note:*** The choice of Z is arbitrary — you can just as easily use the X or Y direction; you can assume that there's some reason to choose the Z direction, such as a magnetic field that exists along that direction.)

If the quantum number of the Z component of the angular momentum is designated by m, then the complete eigenstate is given by $|l, m\rangle$, so you get the following:

$$H|l, m\rangle = \frac{L^2}{2I}|l, m\rangle$$

This chapter takes a look at problems involving angular momentum, including finding the angular momentum eigenvectors and eigenvalues involved with this Hamiltonian.

Rotating Around: Getting All Angular

This section brings you up to speed on angular momentum, taking as its example a disk that's in rotation. Because the disk has mass and is rotating, it has angular momentum. Figure 4-1 shows an example of a rotating disk that illustrates angular momentum.

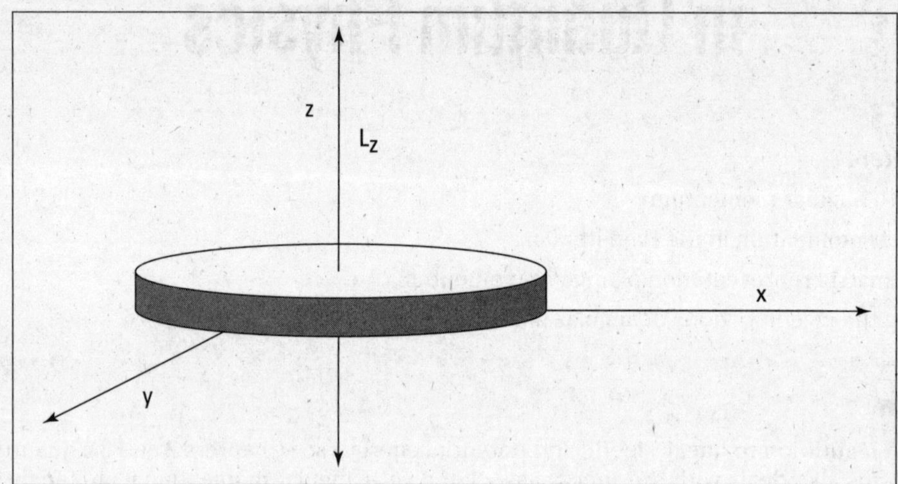

Figure 4-1: A rotating disk.

The disk's angular momentum vector, **L,** is perpendicular to the plane of rotation. If you wrap your right hand in the direction something is rotating, your right thumb points in the direction of the **L** vector.

L is a vector in three-dimensional space, which means that it can point anywhere, which means that it has *x, y,* and *z* components: L_x, L_y, and L_z. Note that **L** is the vector product of **R** and **P** (**L** = **R** × **P**) — that is, position times momentum. You can also write L_x, L_y, and L_z as the following:

- $L_x = YP_z - ZP_y$
- $L_y = ZP_x - XP_z$
- $L_z = XP_y - YP_x$

Here, P_x, P_y, and P_z are the momentum operators, returning the momentum in the *x, y,* and *z* directions, and X, Y, and Z are the position operators, which return the position in the *x, y,* and *z* position.

Note that you can also write the momentum operators P_x, P_y, and P_z as

- $P_x = -i\hbar \frac{\partial}{\partial x}$
- $P_y = -i\hbar \frac{\partial}{\partial y}$
- $P_z = -i\hbar \frac{\partial}{\partial z}$

Keep reading for an example and some practice problems.

Q. Write L_x entirely in position space.

A. $\mathbf{L_x = -i\hbar\left(Y\frac{\partial}{\partial z} - Z\frac{\partial}{\partial y}\right)}$. Here's how you solve the problem:

1. **Write the equation for L_x in terms of the position operators.**

   The equation says that

   $$L_x = YP_z - ZP_y$$

2. **Substitute for the position operators and factor the equation.**

   You want to replace the P_z and P_y. You know that

   - $P_y = -i\hbar\frac{\partial}{\partial y}$
   - $P_z = -i\hbar\frac{\partial}{\partial z}$

   So make the substitution and factor out $-i\hbar$. Here's what you get:

   $$L_x = YP_z - ZP_y = -i\hbar\left(Y\frac{\partial}{\partial z} - Z\frac{\partial}{\partial y}\right)$$

1. Write L_y entirely in position space.

Solve It

2. Write L_z entirely in position space.

Solve It

Untangling Things with Commutators

When applying operators to eigenvectors, you usually want to know what the operators' commutators look like. That lets you change the order the operators are applied in, which is how you figure out what the eigenvalues of the eigenvector are. To do so, you figure out the commutator of two operators to get an idea of what happens when you apply those operators in order.

The trick here is to remember that the square of the angular momentum, L^2, is a *scalar* (that is, just a simple number), not a vector, so it commutes with the L_x, L_y, and L_z operators. That is

- $[L^2, L_x] = 0$
- $[L^2, L_y] = 0$
- $[L^2, L_z] = 0$

The story is a little more involved when you're finding commutators like $[L_x, L_y]$. The following example explains how to do so.

Q. What's the commutator of L_x and L_y?

A. $\left[\mathbf{L}_x, \mathbf{L}_y\right] = i\hbar\mathbf{L}_z$.

First write $[L_x, L_y]$ in terms of the momentum and position operators. In the preceding section, I give you the equations $L_x = YP_z - ZP_y$ and $L_y = ZP_x - XP_z$. So making the substitutions, you know that by definition, the following is true:

$$[L_x, L_y] = [YP_z - ZP_y, ZP_x - XP_z]$$

Rewrite the equation using the distributive law:

$$[L_x, L_y] = [YP_z, ZP_x] - [YP_z, XP_z] - [ZP_y, ZP_x] + [ZP_y, XP_z]$$

Group the terms:

$$[L_x, L_y] = Y[P_z, Z]P_x + X[Z, P_z]P_y$$

Substitute in for the two commutators:

$$\left[L_x, L_y\right] = i\hbar\left(XP_y - YP_x\right)$$

You know $XP_y - YP_x = L_z$, so make the substitution to get the final answer:

$$\left[L_x, L_y\right] = i\hbar L_z$$

3. What's the commutator of L_y and L_z?

Solve It

4. What's the commutator of L_z and L_x?

Solve It

Nailing Down the Angular Momentum Eigenvectors

This section focuses on the eigenvectors of angular momentum. You want to find the eigenstates of angular momentum here as preparation for deriving their eigenvalues and determining what values angular momentum can take on a quantum level.

You start by assuming that the eigenvectors look like $|\alpha, \beta\rangle$ and that

$$L^2|\alpha, \beta\rangle = \hbar^2\alpha|\alpha, \beta\rangle$$

In other words, the eigenvalue of L^2 is $\hbar^2\alpha$, where you have yet to solve for α. Also, the eigenvalue of L_z is

$$L_z|\alpha, \beta\rangle = \hbar\beta|\alpha, \beta\rangle$$

To keep going, you need to introduce raising and lowering operators. That way, you can solve for the ground state by, for example, applying the lowering operator to the ground state, setting the result equal to zero, and then solving for the ground state itself.

Here, the raising operator is L_+ and the lowering operator is L_-, and they raise and lower the L_z quantum number. You can define the raising operator, L_+, and the lowering operator, L_-, this way:

- **Raising:** $L_+ = L_x + iL_y$
- **Lowering:** $L_- = L_x - iL_y$

Note that you can get the *x* and *y* components of the angular momentum with the raising and lowering operators:

- $L_x = \frac{1}{2i}(L_+ + L_-)$
- $L_y = \frac{1}{2i}(L_+ - L_-)$

Q. What is L^2 in terms of L_+, L_-, and L_z?

A. $L^2 = \frac{1}{2}(L_+L_- + L_-L_+) + L_z^2$.

You know that $L_+L_- = L_x^2 + L_y^2 + \hbar L_z$. Using the definition of L, this equals

$$L_+L_- = L^2 - L_z^2 + \hbar L_z$$

Isolate L^2 on the left side of the equation:

$$L^2 = L_+L_- + L_z^2 - \hbar L_z$$

You can also derive that $L^2 = L_-L_+ + L_z^2 + \hbar L_z$, so substitute terms in:

$$L^2 = \frac{1}{2}(L_+L_- + L_-L_+) + L_z^2$$

It's also true that $[L^2, L_\pm] = 0$.

5. If $|\alpha, \beta\rangle$ is an angular momentum eigenvector, what is $L_+|\alpha, \beta\rangle$?

Solve It

6. If $|\alpha, \beta\rangle$ is an angular momentum eigenvector, what is $L^2L_+|\alpha, \beta\rangle$?

Solve It

Obtaining the Angular Momentum Eigenvalues

The preceding section shows you how to nail down the eigenvectors for angular momentum. In quantum physics, you also need to know how to find the eigenvalues for angular momentum because they're the actual values that angular momentum is allowed to take for a specific system.

To find the exact eigenvalues of L^2 and L_z, you apply these operators to $|\alpha, \beta\rangle$. The following example shows you how to do so.

Q. What are the eigenvalues L^2 and L_z when you apply these operators to $|\alpha, \beta\rangle$?

A. $\mathbf{L^2|l, m\rangle = l(l+1)\hbar^2|l, m\rangle}$ **and** $\mathbf{L_z|l, m\rangle = \hbar m|l, m\rangle}$**, where you've renamed the maximum possible value of β (**β_{max}**) as *l* and have renamed β as *m*.**

You know that $L^2 - L_z^2 = L_x^2 + L_y^2$, which is a positive number, so $L^2 - L_z^2 \geq 0$. That means that

$$\langle\alpha, \beta|L^2 - L_z^2|\alpha, \beta\rangle \geq 0$$

And substituting in $L^2|\alpha, \beta\rangle = \alpha\hbar^2$ and $L_z|\alpha, \beta\rangle = \beta\hbar$ gives you the following:

$$\langle\alpha, \beta|L^2 - L_z^2|\alpha, \beta\rangle \geq 0$$

$$\alpha\hbar^2 - (\beta\hbar)^2 \geq 0$$

$$\hbar^2(\alpha - \beta^2) \geq 0$$

Solving the inequality for α tells you that

$$\alpha \geq \beta^2$$

So there's a maximum possible value of β, which you call β_{max}.

There must be a state $|\alpha, \beta_{max}\rangle$ such that you can't raise β anymore, which means that if you apply the raising operator, you get zero:

$$L_+|\alpha, \beta_{max}\rangle = 0$$

Applying the lowering operator to this also gives you zero:

$$L_-L_+|\alpha, \beta_{max}\rangle = 0$$

And from the definition of raising and lowering operators, you can say

$$L_-L_+ = L^2 - L_z^2 - \hbar L_z$$

You can substitute for L_-L_+ and get the following:

$$(L^2 - L_z^2 - \hbar L_z)|\alpha, \beta_{max}\rangle = 0$$

Now put in $L^2|\alpha, \beta_{max}\rangle = \alpha\hbar^2$ and $L_z|\alpha, \beta\rangle = \beta_{max}\hbar$. Here's what you get:

$$\alpha\hbar^2 - (\beta_{max}\hbar)^2 - \beta_{max}\hbar^2 = 0$$

$$(\alpha - \beta_{max}^2 - \beta_{max})\hbar^2 = 0$$

Now solve for α:

$$\alpha = \beta_{max}(\beta_{max} + 1)$$

It's usual to rename β_{max} as *l* and β as *m*, so $|\alpha, \beta\rangle$ becomes $|l, m\rangle$; so here are the answers:

- $L^2|l, m\rangle = l(l+1)\hbar^2|l, m\rangle$
- $L_z|l, m\rangle = m\hbar|l, m\rangle$

And this gives you the eigenvalues of L^2 and L_z, the values those quantities can physically take.

7. Derive the allowed values for l and m in the eigenvector $|l, m\rangle$.

Solve It

8. Using the angular momentum Hamiltonian, derive the following relation:

$$E = \frac{l(l+1)\hbar^2}{2\mu r^2}$$

Solve It

Scoping Out the Raising and Lowering Operators' Eigenvalues

In the earlier sections, you use the raising and lowering operators, but you haven't figured out their eigenstates yet. So what values do you get when you raise or lower the eigenvalues of angular momentum?

The raising and lowering operators work like this:

$$L_{\pm}|l, m\rangle = c|l, m \pm 1\rangle$$

You need to determine what the constant *c* is. Doing so lets you apply the raising operator directly. The following example shows you how to do so. Then you can try a couple of example problems.

Q. If $L_{+}|l, m\rangle = c|l, m+1\rangle$, what is *c*?

A. $\mathbf{c = \hbar\left(l(l+1) - m(m+1)\right)^{1/2}}$**, so**

$$\mathbf{L_{+}|l, m\rangle = \hbar\left(l(l+1) - m(m+1)\right)^{1/2}|l, m+1\rangle.}$$

$L_{+}|l, m\rangle$ gives you a new state, and multiplying that new state by its transpose (see Chapter 2) should give you c^2:

$$\left(L_{+}|l, m\rangle\right)^{\dagger} L_{+}|l, m\rangle = c^2$$

To see this, note that

$$\left(L_{+}|l, m\rangle\right)^{\dagger} L_{+}|l, m\rangle = c^2\langle l, m+1|l, m+1\rangle = c^2$$

On the other hand, also note that

$$\left(L_{+}|l, m\rangle\right)^{\dagger} L_{+}|l, m\rangle = \langle l, m|L_{-}L_{+}|l, m\rangle$$

So you have

$$\langle l, m|L_{-}L_{+}|l, m\rangle = c^2$$

Earlier in the chapter, you see that

$$L_{-}L_{+} = L^2 - L_z^{\,2} - \hbar L_z$$

Therefore, you replace $L_{-}L_{+}$ to get the following:

$$\langle l, m|L^2 - L_z^{\,2} - \hbar L_z|l, m\rangle = c^2$$

Find the square root of both sides. That means that *c* is equal to

$$c = \left(\langle l, m|L^2 - L_z^{\,2} - \hbar L_z|l, m\rangle\right)^{1/2}$$

What is $\left(\langle l, m|L^2 - L_z^2 - \hbar L_z|l, m\rangle\right)^{1/2}$? Apply the L^2 and L_z operators to get the following:

$$c = \hbar\left(l\left(l+1\right) - m\left(m+1\right)\right)^{1/2}$$

And that's the eigenvalue of L_+. And because $L_+|l, m\rangle = c|l, m+1\rangle$, you have this relation:

$$L_+|l, m\rangle = \hbar\left(l\left(l+1\right) - m\left(m+1\right)\right)^{1/2}|l, m+1\rangle$$

9. If $L_-|l, m\rangle = c|l, m-1\rangle$, what is *c*?

Solve It

10. What is $L_+|3, 2\rangle$?

Solve It

Treating Angular Momentum with Matrices

When the levels of angular momentum are evenly spaced, you can treat angular momentum with matrices. Doing so makes handling angular momentum easy. For example, say you have a system with angular momentum, with the total angular momentum quantum number $l = 1$. That means that m can take the values –1, 0, and 1. So you can represent the three possible angular momentum states like this:

- $|1, -1\rangle = \begin{vmatrix} 0 \\ 0 \\ 1 \end{vmatrix}$
- $|1, 0\rangle = \begin{vmatrix} 0 \\ 1 \\ 0 \end{vmatrix}$
- $|1, 1\rangle = \begin{vmatrix} 1 \\ 0 \\ 0 \end{vmatrix}$

Okay, that's what the eigenvectors would look like for an $l = 1$ system. Now what would the various operators look like? The following example and the practice problems show you how to figure it out.

Q. In matrix terms, what would the operator L^2 look like for an $l = 1$ system?

A. $$\mathbf{L^2 = 2\hbar^2 \begin{vmatrix} 1 & 0 & 0 \\ 0 & 1 & 0 \\ 0 & 0 & 1 \end{vmatrix}}$$

You can write L^2 this way in matrix form by using an expectation value in every place:

$$L^2 = \begin{vmatrix} \langle 1, 1|L^2|1, 1\rangle & \langle 1, 1|L^2|1, 0\rangle & \langle 1, 1|L^2|1, -1\rangle \\ \langle 1, 0|L^2|1, 1\rangle & \langle 1, 0|L^2|1, 0\rangle & \langle 1, 0|L^2|1, -1\rangle \\ \langle 1, -1|L^2|1, 1\rangle & \langle 1, -1|L^2|1, 0\rangle & \langle 1, -1|L^2|1, -1\rangle \end{vmatrix}$$

You know that $\langle 1, 1|L^2|1, 1\rangle = l(l+1)\hbar^2 = 2\hbar^2$, that $\langle 1, 1|L^2|1, 0\rangle = 0$, that $\langle 1, 0|L^2|1, 0\rangle = 2\hbar^2$, and so on, so this becomes

$$L^2 = \begin{vmatrix} 2\hbar^2 & 0 & 0 \\ 0 & 2\hbar^2 & 0 \\ 0 & 0 & 2\hbar^2 \end{vmatrix}$$

And you can factor out $2\hbar^2$ and write this as

$$L^2 = 2\hbar^2 \begin{vmatrix} 1 & 0 & 0 \\ 0 & 1 & 0 \\ 0 & 0 & 1 \end{vmatrix}$$

All of which is to say that in matrix form, the equation $L^2|1, 1\rangle = 2\hbar^2|1, 1\rangle$ becomes

$$2\hbar^2 \begin{vmatrix} 1 & 0 & 0 \\ 0 & 1 & 0 \\ 0 & 0 & 1 \end{vmatrix} \begin{vmatrix} 1 \\ 0 \\ 0 \end{vmatrix}$$

which means that

$$2\hbar^2 \begin{vmatrix} 1 \\ 0 \\ 0 \end{vmatrix} = 2\hbar^2 \begin{vmatrix} 1 & 0 & 0 \\ 0 & 1 & 0 \\ 0 & 0 & 1 \end{vmatrix} \begin{vmatrix} 1 \\ 0 \\ 0 \end{vmatrix}$$

11. In matrix terms, what would the operator L_+ look like for an $l = 1$ system?

Solve It

12. In matrix terms, what would the operator L_- look like for an $l = 1$ system?

Solve It

13. In matrix terms, what would the operator L_z look like for an $l = 1$ system?

Solve It

14. In matrix terms, what would the operators L_x and L_y look like for an $l = 1$ system?

Solve It

Answers to Problems on Angular Momentum

The following explanations are the answers to the practice questions I present earlier in this chapter. You can see the original questions, the answers in bold, and the answers worked out, step by step.

1. Write L_y entirely in position space. **The answer is**

$$\mathbf{L_y = ZP_x - XP_z = -i\hbar\left(Z\frac{\partial}{\partial x} - X\frac{\partial}{\partial z}\right)}$$

Write the equation for L_y in terms of the position operators:

$$L_y = ZP_x - XP_z$$

Then substitute for the position operators. You know that $P_x = -i\hbar\frac{\partial}{\partial x}$ and $P_z = -i\hbar\frac{\partial}{\partial z}$, so

$$L_y = ZP_x - XP_z = -i\hbar\left(Z\frac{\partial}{\partial x} - X\frac{\partial}{\partial z}\right)$$

2. Write L_z entirely in position space. **The answer is**

$$\mathbf{L_z = XP_y - YP_x = -i\hbar\left(X\frac{\partial}{\partial y} - Y\frac{\partial}{\partial x}\right)}$$

Write the equation for L_z in terms of the position operators:

$$L_z = XP_y - YP_x$$

Now substitute for the position operators. You know that $P_x = -i\hbar\frac{\partial}{\partial x}$ and $P_y = -i\hbar\frac{\partial}{\partial y}$, so make the substitution and factor the equation:

$$L_z = XP_y - YP_x = -i\hbar\left(X\frac{\partial}{\partial y} - Y\frac{\partial}{\partial x}\right)$$

3. What's the commutator of L_y and L_z? $\mathbf{[L_y, L_z] = i\hbar L_x}$.

$L_y = ZP_x - XP_z$ and $L_z = XP_y - YP_x$, so write $[L_y, L_z]$ in terms of the momentum and position operators:

$$[L_y, L_z] = [ZP_x - XP_z, XP_y - YP_x]$$

Regroup on the right side of the equation:

$$[L_y, L_z] = [ZP_x, XP_y] - [ZP_x, YP_x] - [XP_z, XP_y] + [XP_z, YP_x]$$

Factor out Z and Y:

$$[L_y, L_z] = Z[P_x, X]P_y + Y[X, P_x]P_z$$

This equals

$$[L_y, L_z] = i\hbar(YP_z - ZP_y)$$

But $YP_z - ZP_y = L_x$, so

$$[L_y, L_z] = i\hbar L_x$$

4 What's the commutator of L_z and L_x? $\left[\mathbf{L}_z, \mathbf{L}_x\right] = \boldsymbol{i\hbar}\mathbf{L}_y$.

Write $[L_z, L_x]$ in terms of the momentum and position operators. You know that $L_z = XP_y - YP_x$ and that $L_x = [YP_z - ZP_y]$, so

$$[L_z, L_x] = [XP_y - YP_x, YP_z - ZP_y]$$

Regroup this way:

$$[L_z, L_x] = [XP_y, YP_z] - [XP_y, ZP_y] - [YP_x, YP_z] + [YP_x, ZP_y]$$

Factor out X and Z:

$$[L_z, L_x] = X[P_y, Y]P_z + Z[Y, P_y]P_x$$

This equals

$$\left[L_z, L_x\right] = i\hbar\left(ZP_x - XP_z\right)$$

But $ZP_x - XP_z = L_y$, so

$$\left[L_z, L_x\right] = i\hbar L_y$$

5 If $|\alpha, \beta\rangle$ is an angular momentum eigenvector, what is $L_+|\alpha, \beta\rangle$? $\mathbf{L}_+|\boldsymbol{\alpha}, \boldsymbol{\beta}\rangle = \boldsymbol{c}|\boldsymbol{\alpha}, \boldsymbol{\beta}+\mathbf{1}\rangle$**, where *c* is a constant.**

You want to see what $L_+|\alpha, \beta\rangle$ is. Start by applying the L_z operator on it like this:

$$L_zL_+|\alpha, \beta\rangle = ?$$

You can see that $L_zL_+ - L_+L_z = \hbar L_+$, so $L_zL_+ = L_+L_z + \hbar L_+$. Substitute for L_zL_+ to get the following:

$$L_zL_+|\alpha, \beta\rangle = L_+L_z|\alpha, \beta\rangle + \hbar L_+|\alpha, \beta\rangle$$

$L_z|\alpha, \beta\rangle = \hbar\beta$, so do the substitution:

$$\begin{aligned} L_zL_+|\alpha, \beta\rangle &= L_+L_z|\alpha, \beta\rangle + \hbar L_+|\alpha, \beta\rangle \\ &= L_+|\alpha, \beta\rangle\hbar\beta + \hbar L_+|\alpha, \beta\rangle \\ &= \hbar(\beta+1)L_+|\alpha, \beta\rangle \end{aligned}$$

Therefore, you've just proven that the eigenstate $L_+|\alpha, \beta\rangle$ is also an eigenstate of the L_z operator, with an eigenvalue of $(\beta + 1)$:

$$L_+|\alpha, \beta\rangle = c|\alpha, \beta+1\rangle$$

where *c* is a constant. So the L_+ operator has the effect of raising the β quantum number by 1. Similarly, the lowering operator does this:

$$L_-|\alpha, \beta\rangle = c|\alpha, \beta-1\rangle$$

6 If $|\alpha, \beta\rangle$ is an angular momentum eigenvector, what is $L^2L_+|\alpha, \beta\rangle$? $\mathbf{L}^2\mathbf{L}_+|\boldsymbol{\alpha}, \boldsymbol{\beta}\rangle = \boldsymbol{\alpha\hbar}^2\mathbf{L}_+|\boldsymbol{\alpha}, \boldsymbol{\beta}\rangle$.

Because L^2 is a scalar, it commutes with everything. $L^2L_+ - L_+L^2 = 0$, so the following is true:

$$L^2L_+|\alpha, \beta\rangle = L_+L^2|\alpha, \beta\rangle$$

Because $L^2|\alpha, \beta\rangle = \alpha\hbar^2|\alpha, \beta\rangle$, you can make the following substitution:

$$L^2 L_+|\alpha, \beta\rangle = \alpha\hbar^2 L_+|\alpha, \beta\rangle$$

Similarly, the lowering operator, L_-, gives you this:

$$L^2 L_-|\alpha, \beta\rangle = \alpha\hbar^2 L_-|\alpha, \beta\rangle$$

7 Derive the allowed values for l and m in the eigenvector $|l, m\rangle$. **The answers are** $\mathbf{L^2|l, m\rangle = l(l+1)\hbar^2|l, m\rangle}$**, where** $\mathbf{l = 0, \frac{1}{2}, 1, \frac{3}{2}, \ldots}$ **and** $\mathbf{L_z|l, m\rangle = m\hbar|l, m\rangle}$**, where** $\mathbf{-l \leq m \leq l}$**.**

In addition to a β_{max}, there must also be a β_{min} such that when you apply the lowering operator, L_-, you get zero (you can't go any lower than β_{min}). Here's what that idea looks like mathematically:

$$L_-|l, \beta_{min}\rangle = 0$$

You can apply L_+ on this equation as well:

$$L_+L_-|l, \beta_{min}\rangle = 0$$

$L_+L_- = L^2 - L_z^2 + \hbar L_z$, so make the following substitution:

$$\left(L^2 - L_z^2 + \hbar L_z\right)|l, \beta_{min}\rangle = 0$$

which gives you

$$\left(l - \beta_{min}^2 + \beta_{min}\right)\hbar^2 = 0$$

Now solve for l:

$$l - \beta_{min}^2 + \beta_{min} = 0$$

$$l = \beta_{min}^2 - \beta_{min}$$

$$l = \beta_{min}(\beta_{min} - 1)$$

And so you get from the definition of l that

$$\beta_{max} = -\beta_{min}$$

You reach $|\alpha, \beta_{min}\rangle$ by n successive applications of L_- on $|\alpha, \beta_{max}\rangle$, so write the following equation:

$$\beta_{max} = \beta_{min} + n$$

Combining the previous two equations, you get

$$\beta_{max} = \frac{n}{2}$$

Therefore, β_{max} can be either an integer or half an integer (depending on whether n is even or odd).

Because $l = \beta_{max}$, $m = \beta$, and n is a positive number, you see that

$$-l \leq m \leq l$$

So the eigenstates are $|l, m\rangle$; l is the total angular momentum quantum number, and m is the angular momentum along the z-axis quantum number. So you get the following equations:

- $L^2|l, m\rangle = l(l+1)\hbar^2|l, m\rangle$
- $L_z|l, m\rangle = m\hbar|l, m\rangle$ (or, put another way, $-l \le m \le l$)

8 Using the angular momentum Hamiltonian, derive the following relation:

$$E = \frac{l(l+1)\hbar^2}{2\mu r^2}$$

You end with

$$\mathbf{E} = \frac{l(l+1)\hbar^2}{2\mu r^2}$$

At the beginning of the chapter, I note that the Hamiltonian for angular momentum problems is

$$H = \frac{L^2}{2I}$$

where I is the rotational moment of inertia, which is $I = m_1 r_1^2 + m_2 r_2^2$. You can also write this value as $I = \mu r^2$, where $r = |r_1 - r_2|$ and $\mu = \frac{m_1 m_2}{m_1 + m_2}$. Plug in μr^2 for I, and the Hamiltonian works out to be

$$H = \frac{L^2}{2I} = \frac{L^2}{2\mu r^2}$$

Applying the Hamiltonian to the eigenstates $|l, m\rangle$ gives you

$$H|l, m\rangle = \frac{L^2}{2\mu r^2}|l, m\rangle$$

And because $L^2|l, m\rangle = l(l+1)\hbar^2|l, m\rangle$, applying the Hamiltonian gives you

$$H|l, m\rangle = \frac{l(l+1)\hbar^2}{2\mu r^2}|l, m\rangle$$

The Hamiltonian finds allowed energy states, so you know that $H|l, m\rangle = E|l, m\rangle$. Therefore, write the equation in terms of energy:

$$E = \frac{l(l+1)\hbar^2}{2\mu r^2}$$

9 If $L_-|l, m\rangle = c|l, m-1\rangle$, what is c? **The answer is $\mathbf{c = \hbar\left(l(l+1) - m(m-1)\right)^{1/2}}$, so $\mathbf{L_-|l, m\rangle = \hbar\left(l(l+1) - m(m-1)\right)^{1/2}|l, m-1\rangle}$.**

$L_+|l, m\rangle$ gives you a new state, and multiplying that new state by its transpose should give you c^2:

$$\left(L_+|l, m\rangle\right)^\dagger L_+|l, m\rangle = c^2$$

To see that, note that according to the definition of the lowering operator

$$\left(L_-|l, m\rangle\right)^\dagger L_-|l, m\rangle = c^2\langle l, m-1|l, m-1\rangle = c^2$$

On the other hand, also note that using the definition of the transpose

$$\left(L_-|l, m\rangle\right)^\dagger L_-|l, m\rangle = \langle l, m|L_+L_-|l, m\rangle$$

Therefore, you have the following:

$$\langle l, m|L_+L_-|l, m\rangle = c^2$$

Earlier in the chapter, you see that $L_+L_- = L_z^2 + \hbar L_z$, so plug in the value of L_+L_-. Here's what you get:

$$\langle l, m|L^2 - L_z^2 + \hbar L_z|l, m\rangle = c^2$$

Take the square root of both sides. That means that c is equal to

$$c = \left(\langle l, m|L^2 - L_z^2 + \hbar L_z|l, m\rangle\right)^{1/2}$$

So what is $\left(\langle l, m|L^2 - L_z^2 + \hbar L_z|l, m\rangle\right)^{1/2}$? Applying the L^2 and L_z operators gives you the following:

$$c = \hbar\left(l(l+1) - m(m-1)\right)^{1/2}$$

And that's the eigenvalue of L_+, which means you have this relation:

$$L_-|l, m\rangle = \hbar\left(l(l+1) - m(m-1)\right)^{1/2}|l, m-1\rangle$$

10 What is $L_+|3, 2\rangle$? $\mathbf{L_+|3, 2\rangle = \hbar(6)^{1/2}|3, 3\rangle}$.

You know that

$$L_+|l, m\rangle = \hbar\left(l(l+1) - m(m+1)\right)^{1/2}|l, m+1\rangle$$

Here, $l = 3$ and $m = 2$, so plug in the numbers and perform the calculations:

$$\begin{aligned} L_+|3, 2\rangle &= \hbar\left(3(3+1) - 2(2+1)\right)^{1/2}|3, 3\rangle \\ &= \hbar(12-6)^{1/2}|3, 3\rangle \\ &= \hbar(6)^{1/2}|3, 3\rangle \end{aligned}$$

11 In matrix terms, what would the operator L_+ look like for an $l = 1$ system? **Here's the answer:**

$$\mathbf{L_+} = \sqrt{2}\hbar \begin{vmatrix} 0 & 1 & 0 \\ 0 & 0 & 1 \\ 0 & 0 & 0 \end{vmatrix}$$

Use the raising operator to get the following:

$$L_+|l, m\rangle = \hbar\left(l(l+1) - m(m+1)\right)^{1/2}|l, m+1\rangle$$

Here, $l = 1$, and $m = 1, 0$, and -1. So you have the following:

- $L_+|1, 1\rangle = 0$
- $L_+|1, 0\rangle = \sqrt{2}\hbar|1, 1\rangle$
- $L_+|1, -1\rangle = \sqrt{2}\hbar|1, 0\rangle$

Therefore, the L_+ operator looks like this in matrix form:

$$L_+ = \sqrt{2}\hbar\begin{vmatrix} 0 & 1 & 0 \\ 0 & 0 & 1 \\ 0 & 0 & 0 \end{vmatrix}$$

Note that $L_+|1, 0\rangle$ would be

$$L_+|1, 0\rangle = \sqrt{2}\hbar\begin{vmatrix} 0 & 1 & 0 \\ 0 & 0 & 1 \\ 0 & 0 & 0 \end{vmatrix}\begin{vmatrix} 0 \\ 1 \\ 0 \end{vmatrix}$$

And this equals

$$L_+|1, 0\rangle = \sqrt{2}\hbar\begin{vmatrix} 1 \\ 0 \\ 0 \end{vmatrix} = \sqrt{2}\hbar\begin{vmatrix} 0 & 1 & 0 \\ 0 & 0 & 1 \\ 0 & 0 & 0 \end{vmatrix}\begin{vmatrix} 0 \\ 1 \\ 0 \end{vmatrix}$$

So $\sqrt{2}\hbar|1, 1\rangle = L_+|1, 0\rangle$.

12 In matrix terms, what would the operator L_- look like for an $l = 1$ system? **Here's the answer:**

$$\mathbf{L_- = \sqrt{2}\hbar\begin{vmatrix} 0 & 0 & 0 \\ 1 & 0 & 0 \\ 0 & 1 & 0 \end{vmatrix}}$$

You know that

$$L_-|l, m\rangle = \hbar\left(1(1+1) - m(m-1)\right)^{1/2}|l, m-1\rangle$$

Here, $l = 1$ and $m = 1, 0$, and -1. So that means

- $L_-|1, 1\rangle = \sqrt{2}\hbar|1, 0\rangle$
- $L_-|1, 0\rangle = \sqrt{2}\hbar|1, -1\rangle$
- $L_-|1, -1\rangle = 0$

In other words, the L_- operator looks like this in matrix form:

$$L_- = \sqrt{2}\hbar\begin{vmatrix} 0 & 0 & 0 \\ 1 & 0 & 0 \\ 0 & 1 & 0 \end{vmatrix}$$

That means that $L_-|1, 1\rangle$ would be

$$L_-|1, 1\rangle = \sqrt{2}\hbar\begin{vmatrix} 0 & 0 & 0 \\ 1 & 0 & 0 \\ 0 & 1 & 0 \end{vmatrix}\begin{vmatrix} 1 \\ 0 \\ 0 \end{vmatrix}$$

which equals

$$L_-|1, 1\rangle = \sqrt{2}\hbar\begin{vmatrix} 0 \\ 1 \\ 0 \end{vmatrix} = \sqrt{2}\hbar\begin{vmatrix} 0 & 0 & 0 \\ 1 & 0 & 0 \\ 0 & 1 & 0 \end{vmatrix}\begin{vmatrix} 1 \\ 0 \\ 0 \end{vmatrix}$$

So $\sqrt{2}\hbar|1, 0\rangle = L_-|1, 1\rangle$.

13 In matrix terms, what would the operator L_z look like for an $l = 1$ system?

$$\mathbf{L_z = \hbar\begin{vmatrix} 1 & 0 & 0 \\ 0 & 0 & 0 \\ 0 & 0 & -1 \end{vmatrix}}$$

You know that

- $\hbar|1, 1\rangle = L_z|1, 1\rangle$
- $0 = L_z|1, 0\rangle$
- $-\hbar|1, 1\rangle = L_z|1, -1\rangle$

So that means

$$L_z = \hbar\begin{vmatrix} 1 & 0 & 0 \\ 0 & 0 & 0 \\ 0 & 0 & -1 \end{vmatrix}$$

For example, $L_z|1, -1\rangle$ equals

$$L_z|1, -1\rangle = \hbar\begin{vmatrix} 1 & 0 & 0 \\ 0 & 0 & 0 \\ 0 & 0 & -1 \end{vmatrix}\begin{vmatrix} 0 \\ 0 \\ 1 \end{vmatrix}$$

And this equals

$$L_z|1, -1\rangle = \hbar\begin{vmatrix} 0 \\ 0 \\ 1 \end{vmatrix} = \hbar\begin{vmatrix} 1 & 0 & 0 \\ 0 & 0 & 0 \\ 0 & 0 & -1 \end{vmatrix}\begin{vmatrix} 0 \\ 0 \\ 1 \end{vmatrix}$$

So as you'd expect, $L_z|1, -1\rangle = -\hbar|1, -1\rangle$.

14 In matrix terms, what would the operators L_x and L_y look like for an $l = 1$ system?

$$\mathbf{L}_x = \frac{\hbar}{\sqrt{2}}\begin{vmatrix} 0 & 1 & 0 \\ 1 & 0 & 1 \\ 0 & 1 & 0 \end{vmatrix} \text{ and } \mathbf{L}_y = \frac{\hbar}{\sqrt{2}}\begin{vmatrix} 0 & -i & 0 \\ i & 0 & -i \\ 0 & i & 0 \end{vmatrix}$$

You know that $L_x = \frac{1}{2}(L_+ + L_-)$ and $L_y = \frac{i}{2}(L_- - L_+)$. L_+ equals

$$L_+ = \sqrt{2}\hbar\begin{vmatrix} 0 & 1 & 0 \\ 0 & 0 & 1 \\ 0 & 0 & 0 \end{vmatrix}$$

And L_- equals

$$L_- = \sqrt{2}\hbar\begin{vmatrix} 0 & 0 & 0 \\ 1 & 0 & 0 \\ 0 & 1 & 0 \end{vmatrix}$$

So $L_x = \frac{L_+ + L_-}{2}$ equals

$$L_x = \frac{\hbar}{\sqrt{2}}\begin{vmatrix} 0 & 1 & 0 \\ 1 & 0 & 1 \\ 0 & 1 & 0 \end{vmatrix}$$

And $L_y = \frac{i(L_- - L_+)}{2}$ is

$$L_y = \frac{\hbar}{\sqrt{2}}\begin{vmatrix} 0 & -i & 0 \\ i & 0 & -i \\ 0 & i & 0 \end{vmatrix}$$

What's the commutator, $[L_x, L_y] = L_xL_y - L_yL_x$, look like? First find L_xL_y:

$$L_xL_y = \frac{\hbar^2}{2}\begin{vmatrix} 0 & 1 & 0 \\ 1 & 0 & 1 \\ 0 & 1 & 0 \end{vmatrix}\begin{vmatrix} 0 & -i & 0 \\ i & 0 & -i \\ 0 & i & 0 \end{vmatrix}$$

This equals

$$L_xL_y = \frac{\hbar^2}{2}\begin{vmatrix} 0 & 1 & 0 \\ 1 & 0 & 1 \\ 0 & 1 & 0 \end{vmatrix}\begin{vmatrix} 0 & -i & 0 \\ i & 0 & -i \\ 0 & i & 0 \end{vmatrix} = \frac{\hbar^2}{2}\begin{vmatrix} i & 0 & -i \\ 0 & 0 & 0 \\ i & 0 & -i \end{vmatrix}$$

And similarly, L_yL_x equals

$$L_yL_x = \frac{\hbar^2}{2}\begin{vmatrix} 0 & -i & 0 \\ i & 0 & -i \\ 0 & i & 0 \end{vmatrix}\begin{vmatrix} 0 & 1 & 0 \\ 1 & 0 & 1 \\ 0 & 1 & 0 \end{vmatrix}$$

This equals

$$L_yL_x = \frac{\hbar^2}{2}\begin{vmatrix} 0 & -i & 0 \\ i & 0 & -i \\ 0 & i & 0 \end{vmatrix}\begin{vmatrix} 0 & 1 & 0 \\ 1 & 0 & 1 \\ 0 & 1 & 0 \end{vmatrix} = \begin{vmatrix} -i & 0 & -i \\ 0 & 0 & 0 \\ i & 0 & i \end{vmatrix}$$

So

$$\left[L_x, L_y\right] = L_xL_y - L_yL_x = \frac{\hbar^2}{2}\begin{vmatrix} 2i & 0 & 0 \\ 0 & 0 & 0 \\ 0 & 0 & -2i \end{vmatrix}$$

And this equals

$$\left[L_x, L_y\right] = L_xL_y - L_yL_x = i\hbar^2\begin{vmatrix} 1 & 0 & 0 \\ 0 & 0 & 0 \\ 0 & 0 & -1 \end{vmatrix}$$

Chapter 5

Spin Makes the Particle Go Round

In This Chapter

- Finding spin eigenstates
- Writing spin operators
- Describing spin operators in matrix terms

This chapter is all about the inherent spin built into some particles, such as electrons and protons. Spin is an intrinsic quality — it never goes away. Spin has no spatial component, so you can't express spin using differential operators like the ones I discuss in Chapter 4. You can't find wave functions for spin.

So how do you look at spin? You need to use the bra and ket way of looking at spin — that is, using eigenvectors — because bras and kets aren't tied to any specific representation in spatial terms (I introduce bras and kets in Chapter 1). This chapter walks you through plenty of practice opportunities to get a firmer grasp on spin.

Unless you apply very strong magnetic fields to streams of electrons, you're not likely to come across spin in real life. Its primary feature is that it adds another quantum number to particles like electrons and protons, giving them another quantum state to squeeze into. That's important for particles with half-integral spin, the *fermions,* because no two fermions can occupy exactly the same quantum state — and spin doubles the number of quantum states available.

Introducing Spin Eigenstates

Eigenstates represent every possible state of the system. The good news is that if you're already somewhat familiar with eigenstates for angular momentum (which I discuss in Chapter 4), you can apply what you already know to spin eigenstates. You use the same notation for spin eigenstates that you use for angular momentum, with only minor alterations. For spin eigenstates, you use

- A quantum number, s, that indicates the spin along the z-axis (this number corresponds to the magnitude l in angular momentum)
- A total spin quantum number, m or sometimes m_s (which corresponds to the m in angular momentum)

Note that there's no true z-axis built in for spin — you introduce a z-axis when you apply a magnetic field; by convention, the z-axis is taken to be in the direction of the applied magnetic field.

Just as the notation for angular momentum eigenstates is $|l, m\rangle$, the eigenstates of spin are written as $|s, m\rangle$.

For spin, the m quantum number can take the values $-s, -s + 1, \ldots, s - 1, s$. Some particles have integer spin, $s = 0, 1, 2$ and so on. For example, π mesons have spin $s = 0$, photons have $s = 1$, and so forth. Some particles, such as electrons, protons, and neutrons, have half-integer spin, $s = 1/2$. Δ particles, on the other hand, have $s = 3/2$.

Particles with half-integer spin (1/2, 3/2, and so on) are called *fermions,* and particles with integer spins (0, 1, 2, 3, and so on) are called *bosons.*

Q. What are the spin eigenstates of electrons, $s = 1/2$?

A. $|1/2, 1/2\rangle$ **and** $|1/2, -1/2\rangle$.

The m quantum number can take the values $-s, -s + 1, \ldots, s - 1, s$.

1. What are the spin eigenstates of photons, $s = 1$?

Solve It

2. What are the spin eigenstates of Δ particles, $s = 3/2$?

Solve It

Saying Hello to the Spin Operators: Cousins of Angular Momentum

To work with the spin eigenstates, you need to know the spin operators — that is, the raising and lowering operators. You take a look at how such operators work here.

In quantum physics, the spin operators S^2 (spin squared), S_z (spin in the *z* direction), and S_+ and S_- (spin raising and lowering) are defined by analogy with angular momentum operators L^2, L_z, L_+, and L_- (see Chapter 4 on more on the L^2, L_z, L_+, L_- operators). That is, you can simply replace L with S for the most part.

The following example shows how to figure out these types of problems.

Q. The S^2 operator works like the L^2 operator. What are the eigenvalues of S^2?

A. $\mathbf{S^2|s, m\rangle = s(s+1)\hbar^2|s, m\rangle}$.

In Chapter 4, you discover that the L^2 operator gives you the following result when you apply it to an orbital angular momentum eigenstate:

$$L^2|l, m\rangle = l(l+1)\hbar^2|l, m\rangle$$

The S^2 operator works in a similar fashion. Simply replace each *l* in the angular momentum formula with an *s* to create the spin version:

$$S^2|s, m\rangle = s(s+1)\hbar^2|s, m\rangle$$

3. The S_z operator is defined in analogy to the L_z operator. What are the eigenvalues of S_z?

Solve It

4. The S_+ and S_- operators are defined in analogy to the L_+ and L_- operators. What are the eigenvalues of S_+ and S_-?

Solve It

Living in the Matrix: Working with Spin in Terms of Matrices

In this section, you see how to work with spin in terms of matrices. For simplicity, I restrict this discussion to spin ½ particles.

What would the spin eigenstates look like in eigenvector form? The good news is that you have only two possible states (see Chapter 1 for more on matrices):

- The eigenstate $|\frac{1}{2}, \frac{1}{2}\rangle$ is represented as such:

$$|\tfrac{1}{2}, \tfrac{1}{2}\rangle = \begin{vmatrix} 1 \\ 0 \end{vmatrix}$$

- The eigenstate $|\frac{1}{2}, -\frac{1}{2}\rangle$ is represented as such:

$$|\tfrac{1}{2}, -\tfrac{1}{2}\rangle = \begin{vmatrix} 0 \\ 1 \end{vmatrix}$$

Check out the following example and practice problems for clarification.

Q. What does the S^2 operator look like in matrix terms?

A. $\mathbf{S^2 =}$

$$\mathbf{\frac{3}{4}\hbar^2 \begin{vmatrix} 1 & 0 \\ 0 & 1 \end{vmatrix}}$$

The S^2 operator looks like this in matrix terms, where each element stands for the S^2 operator applied to a pair of states like $\langle\frac{1}{2}, \frac{1}{2}|S^2|\frac{1}{2}, \frac{1}{2}\rangle$:

$$S^2 =$$

$$\begin{vmatrix} \langle\frac{1}{2}, \frac{1}{2}|S_z|\frac{1}{2}, \frac{1}{2}\rangle & \langle\frac{1}{2}, \frac{1}{2}|S_z|\frac{1}{2}, -\frac{1}{2}\rangle \\ \langle\frac{1}{2}, -\frac{1}{2}|S_z|\frac{1}{2}, \frac{1}{2}\rangle & \langle\frac{1}{2}, -\frac{1}{2}|S_z|\frac{1}{2}, -\frac{1}{2}\rangle \end{vmatrix}$$

Using the definition of S^2, that works out to be

$$S^2 =$$

$$\begin{vmatrix} \langle\frac{1}{2}, \frac{1}{2}|S^2|\frac{1}{2}, \frac{1}{2}\rangle & \langle\frac{1}{2}, \frac{1}{2}|S^2|\frac{1}{2}, -\frac{1}{2}\rangle \\ \langle\frac{1}{2}, -\frac{1}{2}|S^2|\frac{1}{2}, \frac{1}{2}\rangle & \langle\frac{1}{2}, -\frac{1}{2}|S^2|\frac{1}{2}, -\frac{1}{2}\rangle \end{vmatrix} = \frac{3}{4}\hbar^2 \begin{vmatrix} 1 & 0 \\ 0 & 1 \end{vmatrix}$$

5. What does the S_z operator look like in matrix terms?

Solve It

6. What do the S_+ and S_- operators look like in matrix terms?

Solve It

Answers to Problems on Spin Momentum

This section provides answers and explanations of the practice questions I present earlier in this chapter.

1. What are the spin eigenstates of photons, $s = 1$? **The answers are $|1, 1\rangle$, $|1, 0\rangle$, and $|1, -1\rangle$.**

   The m quantum number can take the values $-s, -s + 1, \ldots, s - 1, s$, so m can be –1, 0, or 1. Therefore, the eigenvectors $|s, m\rangle$ are $|1, 1\rangle$ and $|1, 0\rangle$ and $|1, -1\rangle$.

2. What are the spin eigenstates of delta particles, $s = 3/2$? **The answers are $|3/2, 3/2\rangle$, $|3/2, 1/2\rangle$, $|3/2, -1/2\rangle$, and $|3/2, -3/2\rangle$.**

   The m quantum number can take the values $-s, -s + 1, \ldots, s - 1, s$, so you know the allowed m values are $-3/2, -1/2, 1/2$, and $3/2$. So the eigenstates are $|3/2, 3/2\rangle$ and $|3/2, 1/2\rangle$ and $|3/2, -1/2\rangle$ and $|3/2, -3/2\rangle$.

3. The S_z operator is defined in analogy to the L_z operator. What are the eigenvalues of S_z? **$\mathbf{S_z |s, m\rangle = m\hbar |s, m\rangle}$.**

   The L_z operator gives you this result when you apply it to an orbital angular momentum eigenstate:

   $$L_z |l, m\rangle = m\hbar |l, m\rangle$$

   And by analogy, the S_z operator works the same way. Replace each l with an s to get the answer:

   $$S_z |s, m\rangle = m\hbar |s, m\rangle$$

4. The S_+ and S_- operators are defined in analogy to the L_+ and L_- operator. What are the eigenvalues of S_+ and S_-? **The answers are $\mathbf{S_+ |s, m\rangle = \hbar(s(s+1) - m(m+1))^{1/2} |s, m+1\rangle}$ and $\mathbf{S_- |s, m\rangle = \hbar(s(s+1) - m(m-1))^{1/2} |s, m-1\rangle}$.**

   L_+ and L_- work like this:

   - $L_+ |l, m\rangle = \hbar(l(l+1) - m(m+1))^{1/2} |l, m+1\rangle$
   - $L_- |l, m\rangle = \hbar(l(l+1) - m(m-1))^{1/2} |l, m-1\rangle$

   Replace the l's with s's, and you get the spin raising and lowering operators.

5. What does the S_z operator look like in matrix terms?

   $$\mathbf{S_z} =$$

   $$\frac{\hbar}{2} \begin{vmatrix} 1 & 0 \\ 0 & -1 \end{vmatrix}$$

   You can represent the S_z operator this way, as a matrix of all its values between the various bras and kets:

   $$S_z =$$

   $$\begin{vmatrix} \langle 1/2, 1/2 | S_z | 1/2, 1/2 \rangle & \langle 1/2, 1/2 | S_z | 1/2, -1/2 \rangle \\ \langle 1/2, -1/2 | S_z | 1/2, 1/2 \rangle & \langle 1/2, -1/2 | S_z | 1/2, -1/2 \rangle \end{vmatrix}$$

Following the definition of S_z (which I introduce earlier in the chapter), this works out to be

$$S_z = \begin{vmatrix} \langle \tfrac{1}{2}, \tfrac{1}{2}|S_z|\tfrac{1}{2}, \tfrac{1}{2}\rangle & \langle \tfrac{1}{2}, \tfrac{1}{2}|S_z|\tfrac{1}{2}, -\tfrac{1}{2}\rangle \\ \langle \tfrac{1}{2}, -\tfrac{1}{2}|S_z|\tfrac{1}{2}, \tfrac{1}{2}\rangle & \langle \tfrac{1}{2}, -\tfrac{1}{2}|S_z|\tfrac{1}{2}, -\tfrac{1}{2}\rangle \end{vmatrix} = \frac{\hbar}{2}\begin{vmatrix} 1 & 0 \\ 0 & -1 \end{vmatrix}$$

As an example, check this answer on the eigenstate $|\tfrac{1}{2}, -\tfrac{1}{2}\rangle$. Finding the z component looks like this:

$$S_z|\tfrac{1}{2}, -\tfrac{1}{2}\rangle$$

Put this in matrix terms. You get the following matrix product, using the definition of S_z and converting the ket to an eigenvector (which just lists the possible states):

$$S_z = \frac{\hbar}{2}\begin{vmatrix} 1 & 0 \\ 0 & -1 \end{vmatrix}\begin{vmatrix} 0 \\ 1 \end{vmatrix}$$

Perform matrix multiplication to get the following:

$$\frac{\hbar}{2}\begin{vmatrix} 1 & 0 \\ 0 & -1 \end{vmatrix}\begin{vmatrix} 0 \\ 1 \end{vmatrix} = -\frac{\hbar}{2}\begin{vmatrix} 0 \\ 1 \end{vmatrix}$$

Finally, put it back into ket notation to get

$$S_z|\tfrac{1}{2}, -\tfrac{1}{2}\rangle = -\frac{\hbar}{2}|\tfrac{1}{2}, -\tfrac{1}{2}\rangle$$

6 What do the S_+ and S_- operators look like in matrix terms? **Here are the answers:**

- $\mathbf{S_+} =$

$$\hbar\begin{vmatrix} \mathbf{0} & \mathbf{1} \\ \mathbf{0} & \mathbf{0} \end{vmatrix}$$

- $\mathbf{S_-} =$

$$\hbar\begin{vmatrix} \mathbf{0} & \mathbf{0} \\ \mathbf{1} & \mathbf{0} \end{vmatrix}$$

The S_+ operator raises the z component of the spin, so it looks like this:

$$S_+ = \hbar\begin{vmatrix} 0 & 1 \\ 0 & 0 \end{vmatrix}$$

And the lowering operator lowers the z component of the spin, so it looks like this:

$$S_- = \hbar\begin{vmatrix} 0 & 0 \\ 1 & 0 \end{vmatrix}$$

So for example, you can find out what $S_+\left|\frac{1}{2}, -\frac{1}{2}\right\rangle$ equals. In matrix terms, it looks like

$$\hbar\begin{vmatrix}0 & 1\\0 & 0\end{vmatrix}\begin{vmatrix}0\\1\end{vmatrix}$$

Performing the multiplication gives you

$$\hbar\begin{vmatrix}0 & 1\\0 & 0\end{vmatrix}\begin{vmatrix}0\\1\end{vmatrix} = \hbar\begin{vmatrix}0\\1\end{vmatrix}$$

In ket form, this is $S_+\left|\frac{1}{2}, -\frac{1}{2}\right\rangle = \hbar\left|\frac{1}{2}, \frac{1}{2}\right\rangle$.

Part III
Quantum Physics in Three Dimensions

"I'll finish this equation and then we'll step back to see the bigger picture."

In this part . . .

This part takes you into the third dimension of solving quantum physics problems. Knowing how to handle 3-D problems is essential in quantum physics because the world is three-dimensional (unlike the one-dimensional problems that I address at the start of this book). Here I show you how to handle quantum physics in three-dimensional rectangular and spherical coordinates.

Chapter 6

Solving Problems in Three Dimensions: Cartesian Coordinates

In This Chapter

- Putting the Schrödinger equation in 3-D
- Looking at free particles in the *x, y,* and *z* directions
- Getting physical solutions with wave packets
- Going from square wells to box wells
- Working with 3-D harmonic oscillators

Now you're ready to take your study of quantum physics to three dimensions. The previous chapters are all concerned with quantum physics in one dimension, but you live in a three-dimensional world, so this chapter takes you into the third dimension. (The good news: You don't need 3-D glasses to practice these problems.)

Solving problems in 3-D doesn't have to be a scary prospect. It mostly involves working with problems that you can separate into three independent equations, one for each dimension. This chapter explains this process and shows you how to find the three-dimensional solutions in Cartesian (that is, rectangular) coordinates.

Taking the Schrödinger Equation to Three Dimensions

The Schrödinger equation is $H\psi(x) = E\psi(x)$, and it lets you solve for the wave function of a particle. The H is the Hamiltonian, E is the energy of the system, and $\psi(x)$ is the wave function of a particle in one dimension. When you take it to 3-D, you take a step closer to real life — after all, how many quantum wells are really just one dimension?

Here's what the Schrödinger equation looks like in one dimension:

$$-\frac{\hbar^2}{2m}\frac{\partial^2}{\partial x^2}\psi(x)+V(x)\psi(x)=E\psi(x)$$

To get to the third dimension, include the y and z components. You generalize the Schrödinger equation in three dimensions like this:

$$-\frac{\hbar^2}{2m}\left(\frac{\partial^2}{\partial x^2}+\frac{\partial^2}{\partial y^2}+\frac{\partial^2}{\partial z^2}\right)\psi(x,y,z)+V(x,y,z)\psi(x,y,z)=E\psi(x,y,z)$$

That's quite a long equation to remember. No worries, though — you can simplify this long equation by using the Laplacian operator. This operator helps you recast this into a more compact form. Here's what the Laplacian looks like:

$$\frac{\partial^2}{\partial x^2}+\frac{\partial^2}{\partial y^2}+\frac{\partial^2}{\partial z^2}=\nabla^2$$

So here's the 3-D Schrödinger equation using the Laplacian:

$$-\frac{\hbar^2}{2m}\nabla^2\psi(x,y,z)+V(x,y,z)\psi(x,y,z)=E\psi(x,y,z)$$

Even after simplifying this equation, it can still be quite difficult to solve. That's why most quantum physics in 3-D limits itself to cases where this differential equation can be separated into one-dimensional versions.

Q. Rewrite the Schrödinger equation in the case where $V(x, y, z) = 0$ for a free particle.

A. $(H_x + H_y + H_z)\psi(x, y, z) = E\psi(x, y, z)$**, where**

- $H_x = -\frac{\hbar^2}{2m}\frac{\partial^2}{\partial x^2}$
- $H_y = -\frac{\hbar^2}{2m}\frac{\partial^2}{\partial y^2}$
- $H_z = -\frac{\hbar^2}{2m}\frac{\partial^2}{\partial z^2}$

Start with the Schrödinger equation in three dimensions:

$$-\frac{\hbar^2}{2m}\left(\frac{\partial^2}{\partial x^2}+\frac{\partial^2}{\partial y^2}+\frac{\partial^2}{\partial z^2}\right)\psi(x,y,z)+V(x,y,z)\psi(x,y,z)=E\psi(x,y,z)$$

A *free particle* is just what it sounds like — a particle with no forces on it. For a free particle, $V(x, y, z) = 0$, so the second term of the Schrödinger equation drops out. Having a free particle also means that you can add the kinetic energy in each of the three dimensions to add up to the total energy. In the Schrödinger equation, the total energy equals

$$-\frac{\hbar^2}{2m}\left(\frac{\partial^2}{\partial x^2}+\frac{\partial^2}{\partial y^2}+\frac{\partial^2}{\partial z^2}\right)$$

So rewrite the total energy as the sum of the kinetic energy to get a new version of the Schrödinger equation:

$$(H_x + H_y + H_z)\psi(x, y, z) = E\psi(x, y, z)$$

where $H_x = -\frac{\hbar^2}{2m}\frac{\partial^2}{\partial x^2}$, $H_y = -\frac{\hbar^2}{2m}\frac{\partial^2}{\partial y^2}$, and $H_z = -\frac{\hbar^2}{2m}\frac{\partial^2}{\partial z^2}$.

1. If the potential $V(x, y, z)$ can be written as the sum of three linearly independent functions, $V(x, y, z) = V_x(x) + V_y(y) + V_z(z)$, what form will the wave function $\psi(x, y, z)$ have?

Solve It

2. Assuming the potential can be written as $V(x, y, z) = V_x(x) + V_y(y) + V_z(z)$, break the Schrödinger equation into three equations.

Solve It

Flying Free with Free Particles in 3-D

The next types of problems you solve in 3-D are ones where you solve for the wave function of free particles. As you may expect, a 3-D free particle wave function is much like a one-dimensional wave function, extended to three dimensions.

For a free particle, the potential is always zero, V(*x*) = V(*y*) = V(*z*) = 0, so the Schrödinger equation becomes the following three equations:

- $-\frac{\hbar^2}{2m}\frac{\partial^2}{\partial x^2}X(x)=E_x X(x)$
- $-\frac{\hbar^2}{2m}\frac{\partial^2}{\partial y^2}Y(y)=E_y Y(y)$
- $-\frac{\hbar^2}{2m}\frac{\partial^2}{\partial z^2}Z(z)=E_z Z(z)$

To make things somewhat simpler, you can rewrite these equations in terms of the wave number, *k:*

$$k^2=\frac{2mE}{\hbar^2}$$

Divide both sides of each equation by $-\hbar^2/(2m)$, and you can plug in k^2 on the right side of each equation. Then the Schrödinger equation becomes the following:

- $\frac{\partial^2}{\partial x^2}X(x)=-k_x^2\,X(x)$
- $\frac{\partial^2}{\partial y^2}Y(y)=-k_y^2\,Y(y)$
- $\frac{\partial^2}{\partial z^2}Z(z)=-k_z^2\,Z(z)$

So you've separated the Schrödinger equation into three equations, and through this kind of divide-and-conquer operation, solving that equation becomes much easier. See the example and then try your hand at the practice problems.

Q. Solve the *x* component of the Schrödinger equation up to an arbitrary multiplicative constant:

$$\frac{\partial^2}{\partial x^2}X(x)=-k_x^2\,X(x)$$

A. $\mathbf{X(x)=Ae^{ik_x x}}$**, where** $\mathbf{k_x^2=\frac{2mE_x}{\hbar^2}}$ **and A is a normalization constant that you have yet to solve for.**

The *x* component of the Schrödinger equation in 3-D is

$$\frac{\partial^2}{\partial x^2}X(x)=-k_x^2\,X(x)$$

Solve this differential equation. You get the following result:

$$X(x)=Ae^{ik_x x}$$

where A is a normalization constant that you have yet to solve for.

3. Solve for the y and z components of the Schrödinger equation up to an arbitrary multiplicative constant:

- $\frac{\partial^2}{\partial y^2}Y(y) = -k_y^2\, Y(y)$
- $\frac{\partial^2}{\partial z^2}Z(z) = -k_z^2\, Z(z)$

Solve It

4. What is the total energy of the particle in terms of the wave numbers in all three dimensions, k_x, k_y, and k_z?

Solve It

Getting Physical by Creating Free Wave Packets

A *wave packet* is a superposition of many different wave functions. You assemble the wave function of real particles into wave packets to make sure such wave functions are finite — otherwise, free-particle wave functions are just simple infinite plane waves.

The wave function for a free particle in three dimensions looks like this:

$$\psi(x, y, z) = X(x)Y(y)Z(z)$$

where $X(x)$, $Y(y)$, and $Z(z)$ are independent functions.

The Schrödinger equation looks like this:

- $-\frac{\hbar^2}{2m}\frac{\partial^2}{\partial x^2}X(x) = E_x\, X(x)$
- $-\frac{\hbar^2}{2m}\frac{\partial^2}{\partial y^2}Y(y) = E_y\, Y(y)$
- $-\frac{\hbar^2}{2m}\frac{\partial^2}{\partial z^2}Z(z) = E_z\, Z(z)$

The solution for the wave function, which you can find using information in the preceding section, is as follows:

$$\psi(x, y, z) = Ae^{i(k_x x + k_y y + k_z z)}$$

You can write this as a shortcut (where $\boldsymbol{r}$ is the radius vector):

$$\psi(\boldsymbol{r}) = Ae^{i(\boldsymbol{k}\cdot\boldsymbol{r})}$$

You can show that the constant A equals the following by normalizing the wave packet (check out Chapter 2 for more on the one-dimensional version of A, which is cubed here for three dimensions):

$$A = \frac{1}{(2\pi)^{3/2}}$$

So you can write the wave function as

$$\psi(\boldsymbol{r}) = \frac{1}{(2\pi)^{3/2}}e^{i(\boldsymbol{k}\cdot\boldsymbol{r})}$$

That's the solution to the Schrödinger equation, but there's one problem — it's *unphysical* (it can't be normalized). The following example and practice problems look at this problem and how to fix it.

Q. Show that this wave function is unphysical for a free particle:

$$\psi(\boldsymbol{r}) = \frac{1}{(2\pi)^{3/2}} e^{i(\boldsymbol{k}\cdot\boldsymbol{r})}$$

A. $$\int_{-\infty}^{+\infty} \psi(\boldsymbol{r},t)\psi^*(\boldsymbol{r},t)d^3\boldsymbol{r} = |C|^2 \int_{-\infty}^{+\infty} d^3\boldsymbol{r} \to \infty$$

For a wave function to be physical, you have to be able to normalize it. That means that the integral of its probability has to add up to 1 — in other words, this integral has to be bounded, not infinite:

$$\int_{-\infty}^{+\infty} \psi(\boldsymbol{r},t)\psi^*(\boldsymbol{r},t)d^3r$$

The problem is that the integral is unbounded:

$$\int_{-\infty}^{+\infty} \psi(\boldsymbol{r},t)\psi^*(\boldsymbol{r},t)d^3r = |C|^2 \int_{-\infty}^{+\infty} d^3\boldsymbol{r} \to \infty$$

Therefore, the wave function cannot be normalized and is unphysical.

5. The key to finding the wave function of a free particle in three dimensions is to construct a wave packet. That looks like this, where ϕ_n is the coefficient of each term in this summation:

$$\psi(\boldsymbol{r}) = \frac{1}{(2\pi)^{3/2}} \sum_{n=1}^{\infty} \phi_n e^{i\boldsymbol{k}_n \cdot \boldsymbol{r}}$$

Convert this summation into integral form, therefore creating a wave packet out of a continuum of wave functions.

Solve It

6. Start with the following form of a wave packet (which you derive in problem 5):

$$\psi(\boldsymbol{r}) = \frac{1}{(2\pi)^{3/2}} \int_{-\infty}^{+\infty} \phi(\boldsymbol{k}) e^{i\boldsymbol{k}\cdot\boldsymbol{r}} d^3\boldsymbol{k}$$

Assume that $\phi(\boldsymbol{k})$ is a Gaussian curve of this form

$$\phi(\boldsymbol{k}) = A \exp\left[\frac{-a^2\boldsymbol{k}^2}{4}\right]$$

and solve for the wave packet's wave function by performing the integral.

Solve It

Getting Stuck in a Box Well Potential

Not all particles are free; particles can also get stuck in potential wells. A *stuck* particle simply has energy that's less than the height of the potential walls, so it can't get out of the well. Particles can get stuck in a box potential, like the following (the three-dimensional version of a square well):

- Inside the box, $V(x, y, z) = 0$
- Outside the box, $V(x, y, z) = \infty$

So you have the following, where L_x, L_y, and L_z are the boundaries of the well:

$$V(x, y, z) = \begin{cases} 0 & 0 < x < L_x, 0 < y < L_y, 0 < z < L_z \\ \infty & \text{otherwise} \end{cases}$$

Assuming that you can separate $V(x, y, z)$ into the independent potentials $V_x(x)$ (a function only of x), $V_y(y)$ (a function only of y), and $V_z(z)$ (a function only of z), you get the following:

- $V_x(x) = \begin{cases} 0 & 0 < x < L_x \\ \infty & \text{otherwise} \end{cases}$
- $V_y(y) = \begin{cases} 0 & 0 < y < L_y \\ \infty & \text{otherwise} \end{cases}$
- $V_z(z) = \begin{cases} 0 & 0 < z < L_z \\ \infty & \text{otherwise} \end{cases}$

The next example takes a closer look at this system. Later in this section, you take a look at two more aspects of box potentials: determining energy levels and normalizing the box potential wave functions.

Q. What is the Schrödinger equation for the box well potential?

A. **If you write the wave function like this**

$\psi(x, y, z) = X(x)Y(y)Z(z)$

then the Schrödinger equation becomes

- $-\frac{\hbar^2}{2m}\frac{\partial^2}{\partial x^2}X(x) = E_x X(x) \quad 0 < x < L_x$
- $-\frac{\hbar^2}{2m}\frac{\partial^2}{\partial y^2}Y(y) = E_y Y(y) \quad 0 < y < L_y$
- $-\frac{\hbar^2}{2m}\frac{\partial^2}{\partial z^2}Z(z) = E_z Z(z) \quad 0 < z < L_z$

Write the general 3-D Schrödinger equation like this:

$$-\frac{\hbar^2}{2m}\nabla^2\psi(x, y, z) + V(x, y, z)\psi(x, y, z) = E\psi(x, y, z)$$

Write out the Laplacian (see the earlier section "Taking the Schrödinger Equation to Three Dimensions") to get the following:

$$-\frac{\hbar^2}{2m}\left(\frac{\partial^2}{\partial x^2} + \frac{\partial^2}{\partial y^2} + \frac{\partial^2}{\partial z^2}\right)\psi(x, y, z) + V(x, y, z)\psi(x, y, z) = E\psi(x, y, z)$$

Because the potential is *separable* (that is, it operates independently in different dimensions), you can write $\psi(x, y, z)$ as

$\psi(x, y, z) = X(x)Y(y)Z(z)$

Inside the box well, the potential — $V(x, y, z)$ — equals zero, so the second term of the Schrödinger equation drops out. Therefore, the Schrödinger equation looks like this for *x, y,* and *z:*

- $-\frac{\hbar^2}{2m}\frac{\partial^2}{\partial x^2}X(x) = E_x X(x) \quad 0 < x < L_x$
- $-\frac{\hbar^2}{2m}\frac{\partial^2}{\partial y^2}Y(y) = E_y Y(y) \quad 0 < y < L_y$
- $-\frac{\hbar^2}{2m}\frac{\partial^2}{\partial z^2}Z(z) = E_z Z(z) \quad 0 < z < L_z$

7. Rewrite the Schrödinger equation in terms of the wave number k, where $k^2 = \frac{2mE}{\hbar^2}$.

Solve It

8. Assuming the wave function for the box well potential has the form $\psi(x, y, z) = X(x)Y(y)Z(z)$, use the Schrödinger equation to solve for the x component, $X(x)$, up to arbitrary normalization constants.

Solve It

Box potentials: Finding those energy levels

As with other quantum physical situations, the energy levels of a particle in a box potential are quantized — only certain energy levels are allowed. You can find the allowed energy levels of a particle in a box potential because the wave function is constrained to go to zero at the boundaries of the box:

$$V(x, y, z) = \begin{cases} 0 & 0 < x < L_x,\ 0 < y < L_y,\ 0 < z < L_z \\ \infty & \text{otherwise} \end{cases}$$

This works because you can write the wave function like the following, where D is a constant:

$$\psi(x, y, z) = D \sin(k_x x) \sin(k_y y) \sin(k_z z)$$

This wave function for a particle in a box potential is all sines because it's constrained to go to zero at $x = y = z = 0$. Having to find where $k_x L_x$ will make the sine go to zero constrains k_x, which in turn tells you the energy, because

$$k_x^2 = \frac{2mE_x}{\hbar^2}$$

This works similarly for the energies associated with the y and z directions.

Q. Constrain k_x, k_y, and k_z in terms of L_x, L_y, and L_z for a particle in the following box potential:

$$V(x,y,z)=\begin{cases}0 & 0<x<L_x,\ 0<y<L_y,\ 0<z<L_z\\ \infty & \text{otherwise}\end{cases}$$

A. **The answers are**

- $\boldsymbol{k_x = \frac{n_x\pi}{L_x} \quad n_x = 1, 2, 3, \ldots}$
- $\boldsymbol{k_y = \frac{n_y\pi}{L_y} \quad n_y = 1, 2, 3, \ldots}$
- $\boldsymbol{k_z = \frac{n_z\pi}{L_z} \quad n_z = 1, 2, 3, \ldots}$

Note that the x component of the wave function must be zero at the box's boundaries, so $X(0) = 0$ and $X(L_x) = 0$.

The fact that $\psi(0) = 0$ tells you right away that the wave function must be a sine. And you know that $X(L_x) = A\sin(k_xL_x) = 0$. Because the sine is zero when its argument is a multiple of π, this means that the following is true:

$$k_xL_x = n_x\pi \qquad n_x = 1, 2, 3, \ldots$$

Now solve for k_x to get your answer:

$$k_x = \frac{n_x\pi}{L_x} \quad n_x = 1, 2, 3, \ldots$$

You find the equations for k_y and k_z the same way; therefore, the wave numbers k_x, k_y, and k_z are

- $k_x = \frac{n_x\pi}{L_x} \quad n_x = 1, 2, 3, \ldots$
- $k_y = \frac{n_y\pi}{L_y} \quad n_y = 1, 2, 3, \ldots$
- $k_z = \frac{n_z\pi}{L_z} \quad n_z = 1, 2, 3, \ldots$

for the following box potential:

$$V(x,y,z)=\begin{cases}0 & 0<x<L_x,\ 0<y<L_y,\ 0<z<L_z\\ \infty & \text{otherwise}\end{cases}$$

9. Solve for the energies E_x, E_y, and E_z, the energies associated with movement in the *x*, *y*, and *z* directions.

Solve It

10. What are the total allowed energies for a particle of mass *m* in the following box potential?

$$V(x, y, z) = \begin{cases} 0 & 0 < x < L_x, 0 < y < L_y, 0 < z < L_z \\ \infty & \text{otherwise} \end{cases}$$

Solve It

Back to normal: Normalizing the wave function

Wave functions have to be normalized — that is, the integral of their square has to be 1 when taken over all space, because the probability of the particle's being somewhere in all space is equal to one.

The wave function in a box potential has this form:

$$\psi(x, y, z) = D \sin(k_x x) \sin(k_y y) \sin(k_z z)$$

But what's D? You need to normalize the wave function to find the value of this arbitrary constant. Check out the following example for more info on how to solve this type of problem.

Q. Normalize the x component of the wave function of a particle in the following potential:

$$V(x,y,z)=\begin{cases}0 & 0<x<L_x,\ 0<y<L_y,\ 0<z<L_z\\ \infty & \text{otherwise}\end{cases}$$

A. $\mathbf{X(x)=\left(\frac{2}{L_x}\right)^{1/2}\sin\left(\frac{n_x\pi x}{L_x}\right)}$ $\quad \mathbf{n_x=1, 2, 3, \ldots}$

Start with what the x component of the wave function looks like (***Hint:*** It's the solution to problem 8 from earlier in this chapter):

$$X(x)=A\sin\left(\frac{n_x\pi x}{L_x}\right)$$

To normalize the x component of the wave function, the following must be true:

$$1=\int_0^{L_x}|X(x)|^2\,dx$$

Plug the value of X(x) into the equation to get the following:

$$1=|A|^2\int_0^{L_x}\sin^2\left(\frac{n_x\pi x}{L_x}\right)dx$$

Now take care of the integral by itself. Perform the integral, which gives you

$$\int_0^{L_x}\sin^2\left(\frac{n_x\pi x}{L_x}\right)dx=\frac{L_x}{2}$$

Then replace the integral with $L_x/2$, which gives you

$$1=|A|^2\left(\frac{L_x}{2}\right)$$

And solve for A:

$$A=\left(\frac{2}{L_x}\right)^{1/2}$$

Finally, plug the value of A into the equation for the x component of the wave function, and you have your answer:

$$X(x)=\left(\frac{2}{L_x}\right)^{1/2}\sin\left(\frac{n_x\pi x}{L_x}\right)\quad n_x=1, 2, 3, \ldots$$

11. Extend the previous result for $X(x)$ to find the form of $\psi(x, y, z)$.

Solve It

12. You have the following box potential, where all sides are of equal length, L:

$$V(x, y, z) = \begin{cases} 0 & 0 < x < L_x, 0 < y < L_y, 0 < z < L_z \\ \infty & \text{otherwise} \end{cases}$$

Solve for

a. The allowed energies

b. The wave function

Solve It

Getting in Harmony with 3-D Harmonic Oscillators

This section extends one-dimensional harmonic oscillators into three dimensions. Here, you work with the Schrödinger equation, the wave function, and allowed energy levels. Check out Chapter 3 for how you handle harmonic oscillators in one dimension.

Q. What is the Schrödinger equation for a three-dimensional harmonic oscillator?

A. $$-\frac{\hbar^2}{2m}\left(\frac{\partial^2}{\partial x^2}+\frac{\partial^2}{\partial y^2}+\frac{\partial^2}{\partial z^2}\right)\psi(x,y,z)$$
$$+\left(\frac{1}{2}m\omega_x^2x^2+\frac{1}{2}m\omega_y^2y^2+\frac{1}{2}m\omega_z^2z^2\right)\psi(x,y,z)=E\psi(x,y,z)$$

In three dimensions, the potential looks like this, the sum of the energies in all three dimensions:

$$V(x,y,z)=\frac{1}{2}m\omega_x^2x^2+\frac{1}{2}m\omega_y^2y^2+\frac{1}{2}m\omega_z^2z^2$$

where

- $\omega_x^2=\frac{k_x}{m}$
- $\omega_y^2=\frac{k_y}{m}$
- $\omega_z^2=\frac{k_z}{m}$

In general, the Schrödinger equation looks like this:

$$-\frac{\hbar^2}{2m}\left(\frac{\partial^2}{\partial x^2}+\frac{\partial^2}{\partial y^2}+\frac{\partial^2}{\partial z^2}\right)\psi(x,y,z)+V(x,y,z)\psi(x,y,z)=E\psi(x,y,z)$$

Substitute in for the three-dimensional potential, $V(x, y, z)$, to get the following equation:

$$-\frac{\hbar^2}{2m}\left(\frac{\partial^2}{\partial x^2}+\frac{\partial^2}{\partial y^2}+\frac{\partial^2}{\partial z^2}\right)\psi(x,y,z)+\left(\frac{1}{2}m\omega_x^2x^2+\frac{1}{2}m\omega_y^2y^2+\frac{1}{2}m\omega_z^2z^2\right)\psi(x,y,z)=E\psi(x,y,z)$$

13. What is the wave function of a three-dimensional harmonic oscillator in terms of $x_0 = \left(\frac{\hbar}{m\omega_x}\right)^{1/2}$, $y_0 = \left(\frac{\hbar}{m\omega_y}\right)^{1/2}$, and $z_0 = \left(\frac{\hbar}{m\omega_z}\right)^{1/2}$, where the quantum numbers in each dimension are n_x, n_y, and n_z?

Solve It

14. What are the allowed energy levels of a three-dimensional harmonic oscillator in terms of ω_x, ω_y, and ω_z, where the quantum numbers in each dimension are n_x, n_y, and n_z?

Solve It

Answers to Problems on 3-D Rectangular Coordinates

Here are the answers to the practice questions I present earlier in this chapter.

1 If the potential $V(x, y, z)$ can be written as the sum of three linearly independent functions, $V(x, y, z) = V_x(x) + V_y(y) + V_z(z)$, what form will the wave function $\psi(x, y, z)$ have? **The answer is $(x, y, z) = X(x)Y(y)Z(z)$.**

The Schrödinger equation, which lets you solve for the wave function, looks like this in 3-D:

$$-\frac{\hbar^2}{2m}\nabla^2\psi(x,y,z)+V(x,y,z)\psi(x,y,z)=E\psi(x,y,z)$$

If you can break up the potential $V(x, y, z)$ into three independent functions, one for each dimension — as in $V(x, y, z) = V_x(x) + V_y(y) + V_z(z)$ — then you can express the wave function that solves the Schrödinger equation as the product of three independent functions like this:

$$\psi(x, y, z) = X(x)Y(y)Z(z)$$

where $X(x)$, $Y(y)$, and $Z(z)$ are independent functions.

2 Assuming the potential can be written as $V(x, y, z) = V_x(x) + V_y(y) + V_z(z)$, break the Schrödinger equation into three equations. **Here are your answers:**

- $-\frac{\hbar^2}{2m}\frac{\partial^2}{\partial x^2}X(x)+V(x)X(x)=E_xX(x)$
- $-\frac{\hbar^2}{2m}\frac{\partial^2}{\partial y^2}Y(y)+V(y)Y(y)=E_yY(y)$
- $-\frac{\hbar^2}{2m}\frac{\partial^2}{\partial z^2}Z(z)+V(z)Z(z)=E_zZ(z)$

where the wave function is $\psi(x, y, z) = X(x)Y(y)Z(z)$.

Assuming you can break the potential into three functions — $V(x, y, z) = V_x(x) + V_y(y) + V_z(z)$ — then you can break the Hamiltonian into three separate operators added together:

$$\left(-\frac{\hbar^2}{2m}\frac{\partial^2}{\partial x^2}+V(x)\right)+\left(-\frac{\hbar^2}{2m}\frac{\partial^2}{\partial y^2}+V(y)\right)+\left(-\frac{\hbar^2}{2m}\frac{\partial^2}{\partial z^2}+V(z)\right)=E$$

Here, the total energy, E, is the sum of the *x* component's energy plus the *y* component's energy plus the *z* component's energy:

$$E = E_x + E_y + E_z$$

So you have three independent Schrödinger equations for the three dimensions:

- $-\frac{\hbar^2}{2m}\frac{\partial^2}{\partial x^2}X(x)+V(x)X(x)=E_xX(x)$
- $-\frac{\hbar^2}{2m}\frac{\partial^2}{\partial y^2}Y(y)+V(y)Y(y)=E_yY(y)$
- $-\frac{\hbar^2}{2m}\frac{\partial^2}{\partial z^2}Z(z)+V(z)Z(z)=E_zZ(z)$

where the wave function is $\psi(x, y, z) = X(x)Y(y)Z(z)$.

3 Solve for the y and z components of the Schrödinger equation up to an arbitrary multiplicative constant:

- $\frac{\partial^2}{\partial y^2} Y(y) = -k_y^2\, Y(y)$
- $\frac{\partial^2}{\partial z^2} Z(z) = -k_z^2\, Z(z)$

$Y(y) = Be^{ik_y y}$ and $Z(z) = Ce^{ik_z z}$, where $k_y^2 = \frac{2mE_y}{\hbar^2}$ and $k_z^2 = \frac{2mE_z}{\hbar^2}$ and where B and C are normalization constants that you have yet to solve for.

You need to solve for the y and z components of the Schrödinger equation in terms of an arbitrary constant. The y and z components look like this in 3-D:

- $\frac{\partial^2}{\partial y^2} Y(y) = -k_y^2 Y(y)$
- $\frac{\partial^2}{\partial z^2} Z(z) = -k_z^2 Z(z)$

Solving these differential equations gives you the following:

- $Y(y) = Be^{ik_y y}$
- $Z(z) = Ce^{ik_z z}$

Note that because $\psi(x, y, z) = X(x)Y(y)Z(z)$, you get the following for $\psi(x, y, z)$:

$$\psi(x, y, z) = De^{ik_x x}e^{ik_y y}e^{ik_z z}$$

where D is a normalization constant you have yet to determine.

The exponent here is the dot product of the vectors $\boldsymbol{k}$ and $\boldsymbol{r}$, $\boldsymbol{k} \cdot \boldsymbol{r}$. That is, if the vector $\boldsymbol{a} = (a_x, a_y, a_z)$ in terms of components and the vector $\boldsymbol{b} = (b_x, b_y, b_z)$, then the dot product of $\boldsymbol{a}$ and $\boldsymbol{b}$ is

$$\boldsymbol{a} \cdot \boldsymbol{b} = (a_x b_x, a_y b_y, a_z b_z)$$

So you can rewrite the wave function as

$$\psi(x, y, z) = De^{i\boldsymbol{k} \cdot \boldsymbol{r}}$$

4 What is the total energy of the particle in terms of the wave numbers in all three dimensions, k_x, k_y, and k_z? **The answer is**

$$\mathbf{E} = \frac{\hbar^2}{2m}\left(k_x^2 + k_y^2 + k_z^2\right)$$

You can add up the energy in each of the three dimensions to get the total energy:

$$E = E_x + E_y + E_z$$

The energy of the x component of the wave function is

$$\frac{\hbar^2 k_x^2}{2m} = E_x$$

Then add in the *y* and *z* components:

$$E = \frac{\hbar^2 k_x^2}{2m} + \frac{\hbar^2 k_y^2}{2m} + \frac{\hbar^2 k_z^2}{2m}$$

$$E = \frac{\hbar^2}{2m}\left(k_x^2 + k_y^2 + k_z^2\right)$$

Note that $k_x^2 + k_y^2 + k_z^2$ is the square of the magnitude of k — that is, k^2. So write the equation for the total energy as

$$E = \frac{\hbar^2}{2m}\left(k_x^2 + k_y^2 + k_z^2\right) = \frac{\hbar^2 k^2}{2m}$$

5 The key to finding the wave function of a free particle in three dimensions is to construct a wave packet. That looks like this, where ϕ_n is the coefficient of each term in this summation:

$$\psi(\boldsymbol{r}) = \frac{1}{(2\pi)^{3/2}} \sum_{n=1}^{\infty} \phi_n e^{i\boldsymbol{k}_n \cdot \boldsymbol{r}}$$

Convert this summation into integral form, therefore creating a wave packet out of a continuum of wave functions. **Here's the answer:**

$$\boldsymbol{\psi(r) = \frac{1}{(2\pi)^{3/2}} \int_{-\infty}^{+\infty} \phi(k) e^{ik\cdot r} d^3k}$$

Here's the sum that you want to convert to integral form:

$$\psi(\boldsymbol{r}) = \frac{1}{(2\pi)^{3/2}} \sum_{n=1}^{\infty} \phi_n e^{i\boldsymbol{k}_n \cdot \boldsymbol{r}}$$

Convert the summation over the various *k* values to an integral over *k*. Here's what you get:

$$\psi(\boldsymbol{r}) = \frac{1}{(2\pi)^{3/2}} \int_{-\infty}^{+\infty} \phi(\boldsymbol{k}) e^{i\boldsymbol{k}\cdot\boldsymbol{r}} d^3k$$

6 Start with the following form of a wave packet (which you derive in problem 5):

$$\psi(\boldsymbol{r}) = \frac{1}{(2\pi)^{3/2}} \int_{-\infty}^{+\infty} \phi(\boldsymbol{k}) e^{i\boldsymbol{k}\cdot\boldsymbol{r}} d^3\boldsymbol{k}$$

Assume that $\phi(\boldsymbol{k})$ is a Gaussian curve of this form

$$\phi(\boldsymbol{k}) = A \exp\left[\frac{-a^2 \boldsymbol{k}^2}{4}\right]$$

and solve for the wave packet's wave function by performing the integral. **The answer is**

$$\boldsymbol{\psi(r) = \left[\frac{2}{\pi a^2}\right]^{3/4} \exp\left[\frac{-r^2}{a^2}\right]}$$

The wave function for the wave packet looks like

$$\psi(\boldsymbol{r}) = \frac{1}{(2\pi)^{3/2}} \int_{-\infty}^{+\infty} \phi(\boldsymbol{k}) e^{i\boldsymbol{k}\cdot\boldsymbol{r}} d^3\boldsymbol{k}$$

where $\phi(\boldsymbol{k}) = \mathrm{A}\exp\left[\frac{-a^2\boldsymbol{k}^2}{4}\right]$.

Begin by normalizing $\phi(\boldsymbol{k})$ to determine what A is. Here's how that works:

$$1 = \int_{-\infty}^{+\infty} |\phi(\boldsymbol{k})|^2 d^3\boldsymbol{k}$$

You make the necessary substitution from the previous equation to get

$$1 = |\mathrm{A}|^2 \int_{-\infty}^{+\infty} \exp\left[\frac{-a^2}{2}\boldsymbol{k}^2\right] d^3\boldsymbol{k}$$

Okay, so you need the value of the integral. Performing the integral gives you the following:

$$1 = |\mathrm{A}|^2 \left[\frac{2\pi}{a^2}\right]^{3/2}$$

Then solve for A:

$$\mathrm{A} = \left[\frac{a^2}{2\pi}\right]^{3/4}$$

Plug the value of A back into the wave function, which gives you

$$\psi(\boldsymbol{r}) = \frac{1}{(2\pi)^{3/2}} \left[\frac{a^2}{2\pi}\right]^{3/4} \int_{-\infty}^{+\infty} \exp\left[\frac{-a^2\boldsymbol{k}^2}{4}\right] e^{i\boldsymbol{k}\cdot\boldsymbol{r}} d^3\boldsymbol{k}$$

Finally, perform the integral, which gives you

$$\psi(\boldsymbol{r}) = \left[\frac{2}{\pi a^2}\right]^{3/4} \exp\left[\frac{-\boldsymbol{r}^2}{a^2}\right]$$

And that's what the wave function for a Gaussian wave packet looks like in 3-D.

7 Rewrite the Schrödinger equation in terms of the wave vector $\boldsymbol{k}$, where $k^2 = \frac{2m\mathrm{E}}{\hbar^2}$. **Here are the answers:**

- $\frac{\partial^2}{\partial x^2}\mathrm{X}(x) = -k_x^2\,\mathrm{X}(x) \quad 0 < x < \mathrm{L}_x$
- $\frac{\partial^2}{\partial y^2}\mathrm{Y}(y) = -k_y^2\,\mathrm{Y}(y) \quad 0 < y < \mathrm{L}_y$
- $\frac{\partial^2}{\partial z^2}\mathrm{Z}(z) = -k_z^2\,\mathrm{Z}(z) \quad 0 < z < \mathrm{L}_z$

If you write the wave function like $\psi(x, y, z) = X(x)Y(y)Z(z)$, then the Schrödinger equation for the box well is

- $-\frac{\hbar^2}{2m}\frac{\partial^2}{\partial x^2}X(x) = E_x X(x) \quad 0 < x < L_x$
- $-\frac{\hbar^2}{2m}\frac{\partial^2}{\partial y^2}Y(y) = E_y Y(y) \quad 0 < y < L_y$
- $-\frac{\hbar^2}{2m}\frac{\partial^2}{\partial z^2}Z(z) = E_z Z(z) \quad 0 < z < L_z$

Divide both sides of the each equation by $-\hbar^2/(2m)$, and you can rewrite these equations in terms of the wave vector, $\boldsymbol{k}$, where

$$k^2 = \frac{2mE}{\hbar^2}$$

The Schrödinger equation becomes

- $\frac{\partial^2}{\partial x^2}X(x) = -k_x^2 X(x) \quad 0 < x < L_x$
- $\frac{\partial^2}{\partial y^2}Y(y) = -k_y^2 Y(y) \quad 0 < y < L_y$
- $\frac{\partial^2}{\partial z^2}Z(z) = -k_z^2 Z(z) \quad 0 < z < L_z$

Note that the particle just acts like a free particle inside the box.

8 Assuming the wave function for the box well potential has the form $\psi(x, y, z) = X(x)Y(y)Z(z)$, use the Schrödinger equation to solve for the x component, $X(x)$, up to arbitrary normalization constants. **$X(x) = A \sin(k_x x) + B \cos(k_x x)$.**

The Schrödinger equation for the x direction for the box well looks like this:

$$\frac{\partial^2}{\partial x^2}X(x) = -k_x^2 X(x) \quad 0 < x < L_x$$

Solve the differential equation, which gives you

- $X_1(x) = A \sin(k_x x)$
- $X_2(x) = B \cos(k_x x)$

where A and B are constants that are yet to be determined. So the solution is the sum of $X_1(x)$ and $X_2(x)$:

$$X(x) = A \sin(k_x x) + B \cos(k_x x)$$

9 Solve for the energies E_x, E_y, and E_z, the energies associated with movement in the *x, y,* and *z* directions. **Here are the answers:**

- $\mathbf{E_x = \dfrac{n_x^2\hbar^2\pi^2}{2mL_x^2} \quad n_x = 1, 2, 3, \ldots}$
- $\mathbf{E_y = \dfrac{n_y^2\hbar^2\pi^2}{2mL_y^2} \quad n_y = 1, 2, 3, \ldots}$
- $\mathbf{E_z = \dfrac{n_z^2\hbar^2\pi^2}{2mL_z^2} \quad n_z = 1, 2, 3, \ldots}$

The problem works the same for the *x, y,* and *z* directions, so you need to do calculations for only one direction. Because $E = \left(n_x + \frac{1}{2}\right)\hbar\omega_x + \left(n_y + \frac{1}{2}\right)\hbar\omega_y + \left(n_z + \frac{1}{2}\right)\hbar\omega_z$, you know that

$$\frac{2mE_x}{\hbar^2} = \frac{n_x^2\pi^2}{L_x^2} \quad n_x = 1, 2, 3, \ldots$$

You want to know the energy, so solve for E_x:

$$E_x = \frac{n_x^2\hbar^2\pi^2}{2mL_x^2} \quad n_x = 1, 2, 3, \ldots$$

Now write the energy for the *y* and *z* directions, which are similarly

- $E_y = \dfrac{n_y^2\hbar^2\pi^2}{2mL_y^2} \quad n_y = 1, 2, 3, \ldots$
- $E_z = \dfrac{n_z^2\hbar^2\pi^2}{2mL_z^2} \quad n_z = 1, 2, 3, \ldots$

10 What are the total allowed energies for a particle of mass *m* in the following box potential?

$$V(x, y, z) = \begin{cases} 0 & 0 < x < L_x,\ 0 < y < L_y,\ 0 < z < L_z \\ \infty & \text{otherwise} \end{cases}$$

And the answers are

$$\mathbf{\frac{n_x^2\hbar^2\pi^2}{2mL_x^2} \quad n_x = 1, 2, 3, \ldots}$$
$$\mathbf{+\frac{n_y^2\hbar^2\pi^2}{2mL_y^2} \quad n_y = 1, 2, 3, \ldots}$$
$$\mathbf{+\frac{n_z^2\hbar^2\pi^2}{2mL_z^2} \quad n_z = 1, 2, 3, \ldots}$$

The total energy of just the particle is just the total energy, $E = E_x + E_y + E_z$, which equals the following:

$$\frac{n_x^2\hbar^2\pi^2}{2mL_x^2} \quad n_x = 1, 2, 3, \ldots$$
$$+\frac{n_y^2\hbar^2\pi^2}{2mL_y^2} \quad n_y = 1, 2, 3, \ldots$$
$$+\frac{n_z^2\hbar^2\pi^2}{2mL_z^2} \quad n_z = 1, 2, 3, \ldots$$

11 Extend the previous result for $X(x)$ to find the form of $\psi(x, y, z)$. **The answer is**

$$\boldsymbol{\psi(x,y,z)=\left(\frac{8}{L_xL_yL_z}\right)^{1/2}\sin\left(\frac{n_x\pi x}{L_x}\right)\sin\left(\frac{n_y\pi y}{L_y}\right)\sin\left(\frac{n_z\pi z}{L_z}\right)}$$

You know the following equation from the example problem, solved for the *x* component of the wave function:

$$X(x)=\left(\frac{2}{L_x}\right)^{1/2}\sin\left(\frac{n_x\pi x}{L_x}\right) \quad n_x = 1, 2, 3, \ldots$$

Solve for $Y(y)$ and $Z(z)$. They work the same way, which gives you

- $Y(y)=\left(\frac{2}{L_y}\right)^{1/2}\sin\left(\frac{n_y\pi x}{L_y}\right) \quad n_y = 1, 2, 3, \ldots$
- $Z(z)=\left(\frac{2}{L_z}\right)^{1/2}\sin\left(\frac{n_z\pi x}{L_z}\right) \quad n_z = 1, 2, 3, \ldots$

You know that

$$\psi(x, y, z) = X(x)Y(y)Z(z)$$

So substitute for the values of $X(x)$, $Y(y)$, $Z(z)$. Here's your answer:

$$\psi(x,y,z)=\left(\frac{8}{L_xL_yL_z}\right)^{1/2}\sin\left(\frac{n_x\pi x}{L}\right)\sin\left(\frac{n_y\pi y}{L}\right)\sin\left(\frac{n_z\pi z}{L}\right)$$
$$n_x = 1, 2, 3, \ldots$$
$$n_y = 1, 2, 3, \ldots$$
$$n_z = 1, 2, 3, \ldots$$

12 You have the following box potential, where all sides are of equal length, L:

$$V(x,y,z)=\begin{cases}0 & 0<x<L_x,\ 0<y<L_y,\ 0<z<L_z\\ \infty & \text{otherwise}\end{cases}$$

Solve for the allowed energies and the wave function. **The allowed energy is**

$$\mathbf{E}=\frac{\hbar^2\pi^2}{2m\mathbf{L}^2}\left(n_x^2+n_y^2+n_z^2\right)$$

And the wave function is $\psi(x,y,z)=\left(\frac{8}{L^3}\right)^{1/2}\sin\left(\frac{n_x\pi x}{L}\right)\sin\left(\frac{n_y\pi y}{L}\right)\sin\left(\frac{n_z\pi z}{L}\right)$

where

$$n_x = 1, 2, 3, \ldots$$

$$n_y = 1, 2, 3, \ldots$$

$$n_z = 1, 2, 3, \ldots$$

a. Solve for the allowed energies. In general, the energy levels of a particle in a box potential are

$$\mathrm{E}=\frac{n_x^2\hbar^2\pi^2}{2m\mathrm{L}_x^2} \quad n_x=1,2,3,\ldots$$

$$+\frac{n_y^2\hbar^2\pi^2}{2m\mathrm{L}_y^2} \quad n_y=1,2,3,\ldots$$

$$+\frac{n_z^2\hbar^2\pi^2}{2m\mathrm{L}_z^2} \quad n_z=1,2,3,\ldots$$

However, when the box is a cube, L_x, L_y, and L_z are equal. Factor the energy equation, and it becomes

$$\mathrm{E}=\frac{\hbar^2\pi^2}{2m\mathrm{L}^2}\left(n_x^2+n_y^2+n_z^2\right)$$

$$n_x=1,2,3,\ldots$$

$$n_y=1,2,3,\ldots$$

$$n_z=1,2,3,\ldots$$

So, for example, the energy of the ground state, where $n_x = n_y = n_z = 1$, is given by this:

$$\mathrm{E}_{111}=\frac{3\hbar^2\pi^2}{2m\mathrm{L}^2}$$

b. Solve for the wave function. The wave function for a cubic potential looks like this:

$$\psi(x,y,z)=\left(\frac{8}{\mathrm{L}^3}\right)^{1/2}\sin\left(\frac{n_x\pi x}{\mathrm{L}}\right)\sin\left(\frac{n_y\pi y}{\mathrm{L}}\right)\sin\left(\frac{n_z\pi z}{\mathrm{L}}\right)$$

$$n_x=1,2,3,\ldots$$

$$n_y=1,2,3,\ldots$$

$$n_z=1,2,3,\ldots$$

So, for example, the ground state ($n_x = 1, n_y = 1, n_z = 1$) $\psi_{111}(x, y, z)$ is

$$\psi_{111}(x,y,z)=\left(\frac{8}{\mathrm{L}^3}\right)^{1/2}\sin\left(\frac{\pi x}{\mathrm{L}}\right)\sin\left(\frac{\pi y}{\mathrm{L}}\right)\sin\left(\frac{\pi z}{\mathrm{L}}\right)$$

And here's $\psi_{211}(x, y, z)$:

$$\psi_{211}(x,y,z)=\left(\frac{8}{L^3}\right)^{1/2}\sin\left(\frac{2\pi x}{L}\right)\sin\left(\frac{\pi y}{L}\right)\sin\left(\frac{\pi z}{L}\right)$$

And $\psi_{121}(x, y, z)$:

$$\psi_{121}(x,y,z)=\left(\frac{8}{L^3}\right)^{1/2}\sin\left(\frac{\pi x}{L}\right)\sin\left(\frac{2\pi y}{L}\right)\sin\left(\frac{\pi z}{L}\right)$$

13 What is the wave function of a three-dimensional harmonic oscillator in terms of $x_0=\left(\frac{\hbar}{m\omega_x}\right)^{1/2}$, $y_0=\left(\frac{\hbar}{m\omega_y}\right)^{1/2}$, and $z_0=\left(\frac{\hbar}{m\omega_z}\right)^{1/2}$, where the quantum numbers in each dimension are n_x, n_y, and n_z? **Here's the answer:**

$$\boldsymbol{\psi(x,y,z)=\frac{1}{\pi^{3/4}}\frac{H_{n_x}\left(\frac{x}{x_0}\right)\exp\left(\frac{-x^2}{2x_0^2}\right)}{\left(2^{n_x}n_x!x_0\right)^{1/2}}\frac{H_{n_y}\left(\frac{y}{y_0}\right)\exp\left(\frac{-y^2}{2y_0^2}\right)}{\left(2^{n_y}n_y!y_0\right)^{1/2}}\frac{H_{n_z}\left(\frac{z}{z_0}\right)\exp\left(\frac{-z^2}{2z_0^2}\right)}{\left(2^{n_z}n_z!z_0\right)^{1/2}}}$$

Because you can separate the potential into three dimensions, you can write $\psi(x, y, z)$ as

$$\psi(x, y, z) = X(x)Y(y)Z(z)$$

Therefore, the Schrödinger equation looks like this for x:

$$-\frac{\hbar^2}{2m}\frac{\partial^2}{\partial z^2}X(x)+\frac{1}{2}m\omega_x^2x^2X(x)=E_xX(x)$$

Chapter 3 handles problems of this kind, and the solution looks like this:

$$X(x)=\frac{1}{\pi^{1/2}\left(2^{n_x}n_x!x_0\right)^{1/2}}H_n\left(\frac{x}{x_0}\right)\exp\left(\frac{-x^2}{2x_0^2}\right)$$

where $x_0=\left(\hbar/\left(m\omega_x\right)\right)^{1/2}$ and $n_x = 0, 1, 2$, and so on. The H_n term indicates a Hermite polynomial, and here's what $H_0(x)$ through $H_5(x)$ look like:

- $H_0(x) = 1$
- $H_1(x) = 2x$
- $H_2(x) = 4x^2 - 2$
- $H_3(x) = 8x^3 - 12x$
- $H_4(x) = 16x^4 - 48x^2 + 12$
- $H_5(x) = 32x^5 - 160x^3 + 120x$

To write the wave function, you need to include the y and z components, which work the same way as the x component. The wave function looks like this:

$$\psi(x,y,z)=\frac{1}{\pi^{3/4}}\frac{H_{n_x}\left(\frac{x}{x_0}\right)\exp\left(\frac{-x^2}{2x_0^2}\right)}{\left(2^{n_x}n_x!x_0\right)^{1/2}}\frac{H_{n_y}\left(\frac{y}{y_0}\right)\exp\left(\frac{-y^2}{2y_0^2}\right)}{\left(2^{n_y}n_y!y_0\right)^{1/2}}\frac{H_{n_z}\left(\frac{z}{z_0}\right)\exp\left(\frac{-z^2}{2z_0^2}\right)}{\left(2^{n_z}n_z!z_0\right)^{1/2}}$$

14 What are the allowed energy levels of a three-dimensional harmonic oscillator in terms of ω_x, ω_y, and ω_z, where the quantum numbers in each dimension are n_x, n_y, and n_z? **The answer is**

$$\mathbf{E=\left(n_x+\frac{1}{2}\right)\hbar\omega_x+\left(n_y+\frac{1}{2}\right)\hbar\omega_y+\left(n_z+\frac{1}{2}\right)\hbar\omega_z}$$

The energy of a one-dimensional harmonic oscillator is

$$E=\left(n+\frac{1}{2}\right)\hbar\omega$$

Now account for the *x*, *y*, and *z* directions. By extension, the energy of a 3-D harmonic oscillator is given by

$$E=\left(n_x+\frac{1}{2}\right)\hbar\omega_x+\left(n_y+\frac{1}{2}\right)\hbar\omega_y+\left(n_z+\frac{1}{2}\right)\hbar\omega_z$$

Note that if you have an isotropic harmonic oscillator, where $\omega_x = \omega_y = \omega_z = \omega$, the energy levels look like this:

$$E=\left(n_x+n_y+n_z+\frac{3}{2}\right)\hbar\omega$$

Chapter 7

Going Circular in Three Dimensions: Spherical Coordinates

In This Chapter

- Solving problems in spherical coordinates
- Looking at free particles in spherical coordinates
- Working with spherical wells
- Doing problems with isotropic harmonic oscillators

Some problems are built for rectangular coordinates, like the ones I discuss in Chapter 6. In those cases, you can look at potential and energy in the x, y, and z directions, and you can break the wave function into parts that correlate to those three dimensions. However, some problems in the 3-D world are set up for spherical coordinates instead. *Spherical coordinates* involve two angles and a radius vector.

A *spherical well* is a potential well that's a sphere, and it looks something like this:

$$V(r)=\begin{cases}0 & 0<r<a\\ \infty & \text{otherwise}\end{cases}$$

In this case, you have a potential that's spherical, and it extends to a certain radius from the origin. Clearly, trying to solve for the wave function here in terms of rectangular coordinates would be very, very difficult. Instead, you use the spherical coordinate system you see in Figure 7-1. Notice how the spherical coordinate system, which uses the coordinates r, θ, and ϕ, compares to corresponding rectangular coordinates x, y, and z.

You should be able to solve problems set up in terms of spherical coordinates — using spherical coordinates. That's what this chapter is all about.

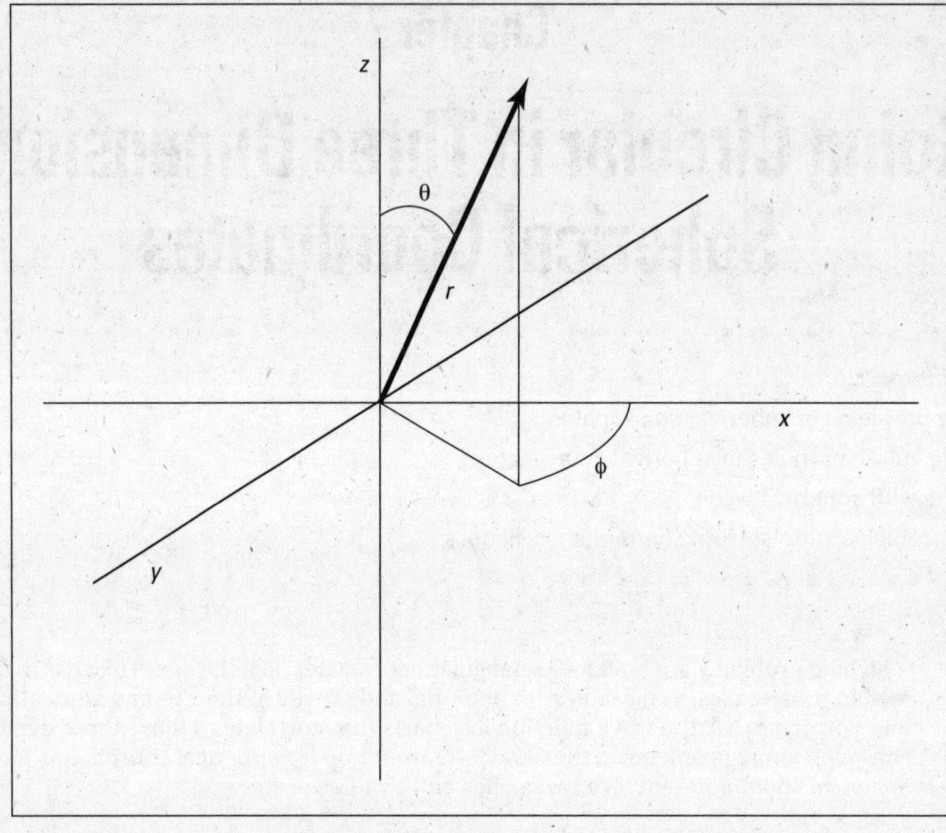

Figure 7-1: The spherical coordinate system.

Taking It to Three Dimensions with Spherical Coordinates

To work in spherical coordinates, you need to get a form of the Schrödinger equation in spherical coordinates. The Schrödinger equation is the basic formula of quantum physics because it lets you solve for the wave functions of particles, and from there you can get all you want — probabilities, expectation values of observables like angular momentum, and so on.

In general, the Schrödinger equation looks like this in spherical coordinates:

$$\frac{-\hbar^2}{2m}\nabla^2\psi(\boldsymbol{r})+V(\boldsymbol{r})\psi(\boldsymbol{r})=E\psi(\boldsymbol{r})$$

where $\boldsymbol{r}$ is the radius vector and ∇^2 is the Laplacian operator.

In spherical coordinates, the Laplacian looks like this:

$$\nabla^2=\frac{1}{r}\frac{\partial^2}{\partial r^2}r-\frac{1}{\hbar^2 r^2}\mathbf{L}^2$$

where $\mathbf{L}^2$ is the square of the orbital angular momentum:

$$\mathbf{L}^2 = -\hbar^2\left(\frac{1}{\sin\theta}\frac{\partial}{\partial\theta}\left(\sin\theta\frac{\partial}{\partial\theta}\right)+\frac{1}{\sin^2\theta}\frac{\partial^2}{\partial\phi^2}\right)$$

All this means that in spherical coordinates, the Schrödinger equation for a central potential looks like this when you substitute in terms:

$$\frac{-\hbar^2}{2m}\frac{1}{r}\frac{\partial^2}{\partial r^2}r\psi(\boldsymbol{r})+\frac{1}{2mr^2}\mathbf{L}^2\psi(\boldsymbol{r})+V(\boldsymbol{r})\psi(\boldsymbol{r})=E\psi(\boldsymbol{r})$$

Note that this is the sum of the radial kinetic energy, the angular kinetic energy, and the potential energy.

To dig into the details, expand $\boldsymbol{r}$ in terms of the r, θ, and ϕ coordinates:

$$\psi(\boldsymbol{r}) = \psi(r, \theta, \phi)$$

For spherically symmetric potentials, you can break the wave function into two parts — a radial part and a part that depends on the angles — like this:

$$\psi(r, \theta, \phi) = R_{nl}(r)\, Y_{lm}(\theta, \phi)$$

In this wave function, here's what the variables mean:

- $R_{nl}(r)$ is the radial part.
- $Y_{lm}(\theta, \phi)$ is the angular part (known as a *spherical harmonic*).
- The n is called the *principal quantum number* (usually associated with energy levels).
- The l is the *total angular momentum quantum number.*
- The m is the *angular momentum quantum number* in the z direction.

The spherical harmonics, $Y_{lm}(\theta, \phi)$, are the eigenfunctions of the $\mathbf{L}^2$ operator. What do those eigenfunctions look like? The following example helps you see how to solve this problem.

Q. What do the functions $Y_{lm}(\theta, \phi)$ look like?

A. $\mathbf{Y}_{lm}(\theta,\phi)=(-1)^m\left(\frac{(2l+1)(l-m)!}{4\pi(l+m)!}\right)^{1/2}\mathbf{P}_{lm}(\cos\theta)e^{im\phi}$ $\quad m \geq 0$, **where**

$\mathbf{P}_{lm}(x)=(1-x^2)^{|m|/2}\frac{d^{|m|}}{dx^{|m|}}\mathbf{P}_l(x)$, **where $\mathbf{P}_l(x)$ is called a Legendre polynomial and is given by**

$$\mathbf{P}_l(x)=\frac{1}{2^l l!}\frac{d^l}{dx^l}(x^2-1)$$

Remember: Solving for $Y_{lm}(\theta, \phi)$ isn't easy, so follow along. Knowing how to deal with spherical harmonics is invaluable when you're finding the wave functions of particles in spherical coordinates, because they're the eigenfunctions of the $\mathbf{L}^2$ operator.

First divide $Y_{lm}(\theta, \phi)$ into a function, $\Theta_{lm}(\theta)$, and an exponential part to get the following:

$$Y_{lm}(\theta, \phi) = \Theta_{lm}(\theta)\frac{e^{im\phi}}{(2\pi)^{1/2}}$$

Then apply the L^2 operator to $Y_{lm}(\theta, \phi)$. Doing so makes $Y_{lm}(\theta, \phi)$ eigenfunctions of the L^2 operator. Here's what you get:

$$\mathbf{L}^2 Y_{lm}(\theta, \phi) = \frac{-\hbar^2}{(2\pi)^{1/2}}\left(\frac{1}{\sin\theta}\frac{\partial}{\partial\theta}\left(\sin\theta\frac{\partial}{\partial\theta}\right) + \frac{1}{\sin^2\theta}\frac{\partial^2}{\partial\phi^2}\right)\Theta_{lm}(\theta)e^{im\phi}$$

Because you're creating $Y_{lm}(\theta, \phi)$ to be an eigenfunction of the $\mathbf{L}^2$ operator, you have this:

$$\begin{aligned}\mathbf{L}^2 Y_{lm}(\theta, \phi) &= l(l+1)\hbar^2 Y_{lm}(\theta, \phi)\\ &= l(l+1)\hbar^2\Theta_{lm}(\theta)\Phi_m(\phi)\end{aligned}$$

Therefore, the previous equation for $L^2Y_{lm}(\theta, \phi)$ becomes

$$\frac{-\hbar^2}{(2\pi)^{1/2}}\left(\frac{1}{\sin\theta}\frac{\partial}{\partial\theta}\left(\sin\theta\frac{\partial}{\partial\theta}\right) + \frac{1}{\sin^2\theta}\frac{\partial^2}{\partial\phi^2}\right)\Theta_{lm}(\theta)e^{im\phi} = l(l+1)\hbar^2\Theta_{lm}(\theta)\frac{e^{im\phi}}{(2\pi)^{1/2}}$$

Cancel terms and subtract the right-hand side from the left, which gives you this differential equation:

$$\left(\frac{1}{\sin\theta}\frac{\partial}{\partial\theta}\left(\sin\theta\frac{\partial}{\partial\theta}\right) + \frac{1}{\sin^2\theta}\frac{\partial^2}{\partial\phi^2}\right)\Theta_{lm}(\theta)e^{im\phi} + l(l+1)\Theta_{lm}(\theta)e^{im\phi} = 0$$

Divide by $e^{im\phi}$ to get the following:

$$\frac{1}{\sin\theta}\frac{\partial}{\partial\theta}\left(\sin\theta\frac{\partial}{\partial\theta}\Theta_{lm}(\theta)\right) + \left(l(l+1) - \frac{m^2}{\sin^2\theta}\right)\Theta_{lm}(\theta) = 0$$

This is a well-known differential equation known as a *Legendre differential equation*. (Check out *Differential Equations For Dummies* [Wiley] for more information.) The solutions are well known and take the form:

$$\Theta_{lm}(\theta) = C_{lm}P_{lm}(\cos\theta)$$

where $P_{lm}(\cos\theta)$ is the *Lengendre function:*

$$P_{lm}(x) = \left(1 - x^2\right)^{|m|/2}\frac{d^{|m|}}{dx^{|m|}}P_l(x)$$

where $P_l(x)$ is called a *Legendre polynomial* and is given by

$$P_l(x) = \frac{1}{2^l l!}\frac{d^l}{dx^l}\left(x^2 - 1\right)^l$$

That gives you $\Theta_{lm}(\theta)$ up to a constant, C_{lm}. And because

$$Y_{lm}(\theta, \phi) = \Theta_{lm}(\theta)\frac{e^{im\phi}}{(2\pi)^{1/2}}$$

you also know what $Y_{lm}(\theta, \phi)$ is — up to a multiplicative constant.

To find the multiplicative constant, you have to normalize $Y_{lm}(\theta, \phi)$; therefore, the following equation must be true:

$$\int_0^{2\pi} d\phi \int_0^{\pi} Y_{lm}^*(\theta,\phi) Y_{lm}(\theta,\phi) \sin\theta \, d\theta = 1$$

Substitute the following equations into the integral:

- $Y_{lm}(\theta,\phi) = \Theta_{lm}(\theta)\Phi_m(\phi)$
- $\Phi_m(\phi) = \dfrac{e^{im\phi}}{(2\pi)^{1/2}}$
- $\Theta_{lm}(\theta) = C_{lm} P_{lm}(\cos\theta)$

The substitution gives you

$$\frac{|C_{lm}|^2}{2\pi} \int_0^{2\pi} d\phi \int_0^{\pi} |P_{lm}^*(\cos\theta)|^2 \sin\theta \, d\theta = 1$$

Doing the integral over ϕ gives you 2π, so this becomes

$$|C_{lm}|^2 \int_0^{\pi} |P_{lm}^*(\cos\theta)|^2 \sin\theta \, d\theta = 1$$

You can do the integral to get

$$|C_{lm}|^2 \frac{2}{2l+1} \frac{(l+m)!}{(l-m)!} = 1$$

Solve for C_{lm}:

$$C_{lm} = (-1)^m \left(\frac{(2l+1)(l-m)!}{2(l+m)!} \right)^{1/2} \quad m \geq 0$$

So plug the value of C_{lm} into the equation $\Theta_{lm}(\theta) = C_{lm}P_{lm}(\cos\theta)$:

$$\Theta_{lm}(\theta) = (-1)^m \left(\frac{(2l+1)(l-m)!}{2(l+m)!} \right)^{1/2} P_{lm}(\cos\theta) \quad m \geq 0$$

And because $Y_{lm}(\theta, \phi) = \Theta_{lm}(\theta)\Phi_m(\phi)$, you plug in the values of $\Theta_{lm}(\theta)$ and $\Phi_m(\phi)$ to get

$$Y_{lm}(\theta,\phi) = (-1)^m \left(\frac{(2l+1)(l-m)!}{4\pi(l+m)!} \right)^{1/2} P_{lm}(\cos\theta)\, e^{im\phi} \quad m \geq 0$$

where $P_{lm}(x) = (1-x^2)^{|m|/2} \dfrac{d^{|m|}}{dx^{|m|}} P_l(x)$, where $P_l(x)$ is a Legendre polynomial and is given by

$$P_l(x) = \frac{1}{2^l l!} \frac{d^l}{dx^l} (x^2-1)$$

1. The solution to the Schrödinger equation in spherical coordinates with a spherical potential is $\psi(r, \theta, \phi) = R_{nl}(r)\, Y_{lm}(\theta, \phi)$, where $Y_{lm}(\theta, \phi)$ is a spherical harmonic, l is the total angular momentum quantum number, and m is the angular momentum quantum number in the z direction. Find $Y_{00}(\theta, \phi)$, $Y_{10}(\theta, \phi)$, $Y_{1\pm1}(\theta, \phi)$, $Y_{20}(\theta, \phi)$, $Y_{2\pm1}(\theta, \phi)$, and $Y_{2\pm2}(\theta, \phi)$.

Solve It

---

2. The wave function looks like this in spherical coordinates:

$$\psi(r, \theta, \phi) = R_{nl}(r)\, Y_{lm}(\theta, \phi)$$

And the Schrödinger equation looks like this:

$$-\frac{\hbar^2}{2m}\frac{1}{r}\frac{\partial^2}{\partial r^2} r\,\psi(\mathbf{r}) + \frac{1}{2mr^2}\mathbf{L}^2\psi(\mathbf{r}) + V(r)\psi(\mathbf{r}) = E\psi(\mathbf{r})$$

You've already substituted the wave function into the Schrödinger equation and solved for $Y_{lm}(\theta, \phi)$. Now substitute the wave function into the Schrödinger equation and get an equation for $R_{nl}(r)$. (***Note:*** You can't solve for $R_{nl}(r)$ yet, because you don't know the form of $V(r)$, but you can get the Schrödinger equation for the radial part of the wave function.)

Solve It

Dealing Freely with Free Particles in Spherical Coordinates

In quantum physics, you also encounter free particles in spherical coordinates. *Free particles* are particles with no force on them, so in spherical coordinates, you have to take special care to get things right. What does the wave function look like for a free particle in 3-D spherical coordinates? You know the wave function has this form in general:

$$\psi(r, \theta, \phi) = R_{nl}(r)\, Y_{lm}(\theta, \phi)$$

And you know $Y_{lm}(\theta, \phi)$ is the angular part. But what does the radial part, $R_{nl}(r)$, look like for a free particle? The Schrödinger equation for the radial part of the wave function looks like this:

$$-\frac{\hbar^2}{2m}\frac{\partial^2}{\partial r^2}\left(r\,R_{nl}(r)\right)+\left(V(r)+\frac{l(l+1)\hbar^2}{2mr^2}\right)\left(r\,R_{nl}(r)\right)=E\left(r\,R_{nl}(r)\right)$$

For a free particle, $V(r) = 0$, so the radial equation becomes

$$-\frac{\hbar^2}{2m}\frac{\partial^2}{\partial r^2}\left(r\,R_{nl}(r)\right)+\frac{l(l+1)\hbar^2}{2mr^2}\left(r\,R_{nl}(r)\right)=E\left(r\,R_{nl}(r)\right)$$

People handle this equation by making the substitution $\rho = kr$, where $k=\left(2mE_n\right)^{1/2}/\hbar$, so that $R_{nl}(r)$ becomes $R_l(kr) = R_l(\rho)$. This substitution means that you get

$$\frac{d^2R_l(\rho)}{d\rho^2}+\frac{2}{\rho}\frac{dR_l(\rho)}{d\rho}+\left(1-\frac{l(l+1)}{\rho^2}\right)R_l(\rho)=0$$

The good news: A well-known solution for this equation is a combination of two functions, the spherical Bessel functions, $j_l(\rho)$, and the spherical Neumann functions, $n_l(\rho)$:

$$R_l(\rho) = A_l j_l(\rho) + B_l n_l(\rho)$$

So the solution is a combination of Bessel and Neumann functions. Here's what the Bessel and Neumann functions equal:

- **Spherical Bessel functions:** $j_l(\rho)=(-\rho)^l\left(\frac{1}{\rho}\frac{d}{d\rho}\right)^l\frac{\sin\rho}{\rho}$
- **Spherical Neumann functions:** $n_l(\rho)=-(-\rho)^l\left(\frac{1}{\rho}\frac{d}{d\rho}\right)^l\frac{\cos\rho}{\rho}$

You can make the solutions easier to handle if you take a look at the spherical Bessel functions and Neumann functions for small and large ρ. The following example gives you a clearer picture.

Q. What do the spherical Bessel functions and Neumann functions look like for small and large ρ?

A. **For small ρ, the Bessel functions reduce to**

$$j_l(\rho) \approx \frac{2^l l! \rho^l}{(2l+1)!}$$

And for small ρ, the Neumann functions reduce to

$$n_l(\rho) \approx -\frac{(2l-1)! \rho^{-l-1}}{2^l l!}$$

For large ρ, the Bessel functions reduce to

$$j_l(\rho) \approx \frac{1}{\rho} \sin\left(\rho - \frac{l\pi}{2}\right)$$

For large ρ, the Neumann functions reduce to

$$n_l(\rho) \approx -\frac{1}{\rho} \cos\left(\rho - \frac{l\pi}{2}\right)$$

Start with the spherical Bessel functions:

$$j_l(\rho) = (-\rho)^l \left(\frac{1}{\rho}\frac{d}{d\rho}\right)^l \frac{\sin\rho}{\rho}$$

And start with the spherical Neumann functions:

$$n_l(\rho) = -(-\rho)^l \left(\frac{1}{\rho}\frac{d}{d\rho}\right)^l \frac{\cos\rho}{\rho}$$

Put in small ρ for the Bessel functions to get the following:

$$j_l(\rho) \approx \frac{2^l l! \rho^l}{(2l+1)!}$$

Do the same for the Neumann function. For small ρ, the Neumann function becomes

$$n_l(\rho) \approx \frac{-(2l-1)! \rho^{-l-1}}{2^l l!}$$

Insert the large ρ in the Bessel functions to get

$$j_l(\rho) \approx \frac{1}{\rho} \sin\left(\rho - \frac{l\pi}{2}\right)$$

For large ρ, the Neumann functions become

$$n_l(\rho) \approx -\frac{1}{\rho} \cos\left(\rho - \frac{l\pi}{2}\right)$$

The Neumann functions diverge for small ρ, and that means that any wave function that includes the Neumann functions would also diverge, which is unphysical. Therefore, the Neumann functions are not acceptable functions in the wave function for small ρ. So that means the wave function $\psi(r, \theta, \phi)$, which equals $R_{nl}(r)\, Y_{lm}(\theta, \phi)$, equals this for small ρ:

$$\psi(r, \theta, \phi) = j_l(kr)\, Y_{lm}(\theta, \phi)$$

where $k = (2mE_n)^{1/2}/\hbar$. And you can approximate this as

$$\psi(r, \theta, \phi) \approx \frac{2^l l! \rho^l}{(2l+1)!} Y_{lm}(\theta, \phi)$$

3. The spherical Bessel functions are given by

$$j_l(\rho) = (-\rho)^l \left(\frac{1}{\rho} \frac{d}{d\rho} \right)^l \frac{\sin\rho}{\rho}$$

Find $j_0(\rho)$, $j_1(\rho)$, and $j_2(\rho)$.

Solve It

4. The spherical Neumann functions are given by

$$n_l(\rho) = -(-\rho)^l \left(\frac{1}{\rho} \frac{d}{d\rho} \right)^l \frac{\cos\rho}{\rho}$$

Find $n_0(\rho)$, $n_1(\rho)$, and $n_2(\rho)$.

Solve It

Getting the Goods on Spherical Potential Wells

This section takes a look at particles trapped in spherical potential wells. That is, you look at cases where a particle doesn't have enough energy to escape entrapment by a spherically symmetric well.

Suppose you have a spherical potential well like this:

$$V(r) = \begin{cases} -V_0 & 0 < r < a \\ 0 & r > a \end{cases}$$

This potential is spherically symmetric, and it varies only in r, not in θ or ϕ. That means that the spherical harmonics apply, and you need to solve only for the radial part of the wave function. Check out the following example and the practice problems.

Q. What is the radial part of the wave function for a particle trapped in a spherical potential well (up to arbitrary normalization constants)?

A. **$R_l(\rho) = A_l j_l(\rho) + B_l n_l(\rho)$. That is, the radial part of the wave function is a combination of the spherical Bessel functions ($j_l(\rho)$) and the spherical Neumann functions ($n_l(\rho)$).**

The radial equation looks like the following for the region $0 < r < a$:

$$-\frac{\hbar^2}{2m}\frac{\partial^2}{\partial r^2}\left(r\,R_{nl}(r)\right) + \left(V(r) + \frac{l(l+1)\hbar^2}{2mr^2}\right)\left(r\,R_{nl}(r)\right) = E\left(r\,R_{nl}(r)\right)$$

In this region, $V(r) = -V_0$, so you have

$$-\frac{\hbar^2}{2m}\frac{\partial^2}{\partial r^2}\left(r\,R_{nl}(r)\right) + \left(-V_0 + \frac{l(l+1)\hbar^2}{2mr^2}\right)\left(r\,R_{nl}(r)\right) = E\left(r\,R_{nl}(r)\right)$$

Take the V_0 term over to the right side of the equation. Here's what you get:

$$-\frac{\hbar^2}{2m}\frac{\partial^2}{\partial r^2}\left(r\,R_{nl}(r)\right) + \frac{l(l+1)\hbar^2}{2mr^2}\left(r\,R_{nl}(r)\right) = (E + V_0)\,r\,R_{nl}(r)$$

Divide by r, which gives you

$$-\frac{\hbar^2}{2m}\frac{\partial^2}{\partial r^2}\left(r\,R_{nl}(r)\right) + \frac{l(l+1)\hbar^2}{2mr^2}R_{nl}(r) = (E + V_0)\,R_{nl}(r)$$

Multiply by $-2m/\hbar^2$. You get

$$\frac{1}{r}\frac{\partial^2}{\partial r^2}\left(r\,R_{nl}(r)\right) - \frac{l(l+1)}{r^2}R_{nl}(r) = \frac{-2m}{\hbar^2}(E + V_0)\,R_{nl}(r)$$

Now change the variable so that $\rho = kr$, where $k = \left(2m\left(E + V_0\right)\right)^{1/2} / \hbar$, so that $R_{nl}(r)$ becomes $R_l(kr) = R_l(\rho)$. You get the following:

$$\frac{d^2R_l(\rho)}{d\rho^2} + \frac{2}{\rho}\frac{dR_l(\rho)}{d\rho} + \left(1 - \frac{l(l+1)}{\rho^2}\right)R_l(\rho) = 0$$

This is the spherical Bessel equation, just as you see for the free particle (see the preceding section, "Dealing Freely with Free Particles in Spherical Coordinates") — but this time, $k = \left(2m\left(E + V_0\right)\right)^{1/2} / \hbar$, not $k = \left(2mE\right)^{1/2} / \hbar$.

The solution is a combination of the spherical Bessel functions, $j_l(\rho)$, and the spherical Neumann functions, $n_l(\rho)$:

$$R_l(\rho) = A_l j_l(\rho) + B_l n_l(\rho)$$

5. Using the case where ρ is small, show that B must equal 0 in the radial part of the wave function solution for $0 < r < a$

$$R_l(\rho) = A_l j_l(\rho) + B_l n_l(\rho)$$

giving you this wave function:

$$\psi_{inside}(r, \theta, \phi) = A_l j_l(\rho_{inside})\, Y_{lm}(\theta, \phi)$$

Solve It

6. What's the wave function look like for $r > a$? Use boundary conditions (continuity of the wave function and its first derivative) to set up equations to find the wave function's normalization constants; don't attempt to find the normalization constants unless you have a lot of time on your hands!

Solve It

Bouncing Around with Isotropic Harmonic Oscillators

A *3-D isotropic harmonic oscillator* is one whose potential is spherically symmetric. This section takes a look at finding the wave functions for isotropic harmonic oscillators. In one dimension, the harmonic oscillator potential is written like this:

$$V(x) = \frac{1}{2} m\omega^2 x^2$$

where $\omega^2 = k/m$ and k is the spring constant (the restoring force of the harmonic oscillator is $F = -kx$).

You can convert this into three-dimensional versions of the harmonic potential by replacing x, the displacement in one dimension, with r, the length of the radial vector in the spherical coordinate system:

$$V(r) = \frac{1}{2} m\omega^2 r^2$$

where $\omega^2 = k/m$.

So what does the Schrödinger equation, which gives you the wave functions and the energy levels, look like — and what are the known solutions to it? Check out the following example.

Q. Find the Schrödinger equation for an isotropic harmonic oscillator, and list the solutions for the radial part of the wave functions.

A. $\mathbf{R}_{nl}(r) = \mathbf{C}_{nl} r^l \exp\left(\frac{-m\omega\, r^2}{2\hbar}\right) L_{(n-l)/2}^{(l+1/2)}\left(\frac{m\omega r^2}{\hbar}\right)$, **where exp($x$) = e^x and**

$$\mathbf{C}_{nl} = \frac{\left(\dfrac{2^{n+l+2}\left(\dfrac{m\omega}{\hbar}\right)^{l+3/2}}{\pi^{1/2}}\right)^{1/2}\left[\dfrac{(n-l)}{2}\right]!\left[\dfrac{(n+l)}{2}\right]!}{(n+l+1)!^{1/2}}$$

and the $L_a^b(r)$ functions are the generalized Laguerre polynomials:

$$\mathbf{L}_a^b(r) = \frac{r^{-b} e^r}{a!} \frac{d^a}{dr^a}\left(e^{-r} r^{a+b}\right)$$

Start with the Schrödinger equation in three dimensions:

$$-\frac{\hbar^2}{2m}\frac{\partial^2}{\partial r^2}\left(r\, R_{nl}(r)\right) + \left(V(r) + \frac{l(l+1)\hbar^2}{2mr^2}\right)\left(r\, R_{nl}(r)\right) = E\left(r\, R_{nl}(r)\right)$$

Here, $V(r)$ looks like

$$V(r) = \frac{1}{2} m\omega^2 r^2$$

where $\omega^2 = k/m$.

Substitute the value of $V(r)$ into the Schrödinger equation, which gives you

$$-\frac{\hbar^2}{2m}\frac{\partial^2}{\partial r^2}\left(r\,R_{nl}(r)\right)+\left(\frac{1}{2}m\omega^2 r^2+\frac{l(l+1)\hbar^2}{2mr^2}\right)\left(r\,R_{nl}(r)\right)=E\left(r\,R_{nl}(r)\right)$$

Because this potential is spherically symmetric, the wave function is going to be of the form

$$\psi(r, \theta, \phi) = R_{nl}(r)\, Y_{lm}(\theta, \phi)$$

where you have yet to solve for the radial function, $R_{nl}(r)$, and $Y_{lm}(\theta, \phi)$ represents the spherical harmonics.

The solution to this Schrödinger equation is well known, and here it is, where $R_{nl}(r)$ is the radial part of the wave function, n is the principal quantum number (the energy level), and l is the total angular momentum quantum number:

$$R_{nl}(r)=C_{nl}r^l\exp\left(\frac{-m\omega r^2}{2\hbar}\right)L_{(n-l)/2}^{(l+1)/2}\left(\frac{m\omega r^2}{\hbar}\right)$$

where $\exp(x) = e^x$ and

$$C_{nl}=\frac{\left(\dfrac{2^{n+l+2}\left(\dfrac{m\omega}{\hbar}\right)^{l+3/2}}{\pi^{1/2}}\right)^{1/2}\left[\dfrac{(n-l)}{2}\right]!\left[\dfrac{(n+l)}{2}\right]!}{(n+l+1)!^{1/2}}$$

and the $L_a^b(r)$ functions are the generalized *Laguerre polynomials*:

$$L_a^b(r)=\frac{r^{-b}e^r}{a!}\frac{d^a}{dr^a}\left(e^{-r}r^{a+b}\right)$$

7. The radial part of the wave function for isotropic harmonic oscillators relies on the generalized Laguerre polynomials:

$$L_a^b(r) = \frac{r^{-b}e^r}{a!}\frac{d^a}{dr^a}\left(e^{-r}r^{a+b}\right)$$

Find $L_0^b(r)$, $L_1^b(r)$, $L_2^b(r)$, and $L_3^b(r)$.

Solve It

8. Find the full expressions for $\psi_{110}(r, \theta, \phi)$, $\psi_{11\pm1}(r, \theta, \phi)$, and $\psi_{200}(r, \theta, \phi)$ for a particle of mass *m* in an isotopic harmonic oscillator.

Solve It

Answers to Problems on 3-D Spherical Coordinates

The following are the answers to the practice questions that I present earlier in this chapter. Here you see the original questions, the answers in bold, and step-by-step answer explanations.

1 The solution to the Schrödinger equation in spherical coordinates with a spherical potential is $\psi(r, \theta, \phi) = R_{nl}(r)\, Y_{lm}(\theta, \phi)$, where $Y_{lm}(\theta, \phi)$ is a spherical harmonic, l is the total angular momentum quantum number, and m is the angular momentum quantum number in the z direction. Find $Y_{00}(\theta, \phi)$, $Y_{10}(\theta, \phi)$, $Y_{1\pm1}(\theta, \phi)$, $Y_{20}(\theta, \phi)$, $Y_{2\pm1}(\theta, \phi)$ and $Y_{2\pm2}(\theta, \phi)$. **Here are the answers:**

- $\mathbf{Y_{00}(\theta, \phi) = \frac{1}{(4\pi)^{1/2}}}$
- $\mathbf{Y_{10}(\theta, \phi) = \left(\frac{3}{4\pi}\right)^{1/2} \cos\theta}$
- $\mathbf{Y_{1\pm1}(\theta, \phi) = \mp\left(\frac{3}{8\pi}\right)^{1/2} e^{\pm i\phi} \sin\theta}$
- $\mathbf{Y_{20}(\theta, \phi) = \left(\frac{5}{16\pi}\right)^{1/2} (3\cos^2\theta - 1)}$
- $\mathbf{Y_{2\pm1}(\theta, \phi) = \mp\left(\frac{15}{8\pi}\right)^{1/2} e^{\pm i\phi} \sin\theta}$
- $\mathbf{Y_{2\pm2}(\theta, \phi) = \left(\frac{15}{32\pi}\right)^{1/2} e^{\pm 2i\phi} \sin^2\theta}$

Here's what $Y_{lm}(\theta, \phi)$ equals:

$$Y_{lm}(\theta, \phi) = (-1)^m \left(\frac{(2l+1)(l-m)!}{4\pi(l+m)!} \right)^{1/2} P_{lm}(\cos\theta) e^{im\phi} \quad m \geq 0$$

where $P_{lm}(x) = (1-x^2)^{|m|/2} \frac{d^{|m|}}{dx^{|m|}} P_l(x)$, where $P_l(x)$ is a Legendre polynomial and is given by $P_l(x) = \frac{1}{2^l l!} \frac{d^l}{dx^l}(x^2-1)$.

First, find the Legendre polynomials for $l = 0$, $l = 1$, and $l = 2$:

- $P_0(x) = 1$
- $P_1(x) = x$
- $P_2(x) = \frac{1}{2}(3x^2 - 1)$

Then find the needed Legendre functions by plugging in the answers for the Legendre polynomials and using $m = 0$, $m = 1$, and $m = 2$:

- $P_{10}(x) = x$
- $P_{20}(x) = \frac{1}{2}(3x^2 - 1)$
- $P_{11}(x) = (1 - x^2)^{1/2}$
- $P_{21}(x) = 3x(1 - x^2)^{1/2}$
- $P_{22}(x) = 3(1 - x^2)^{1/2}$

Then use the Legendre functions and the $Y_{lm}(\theta, \phi)$ equation to get the spherical harmonics:

- $Y_{00}(\theta, \phi) = \frac{1}{(4\pi)^{1/2}}$
- $Y_{10}(\theta, \phi) = \left(\frac{3}{4\pi}\right)^{1/2} \cos\theta$
- $Y_{1\pm1}(\theta, \phi) = \mp\left(\frac{3}{8\pi}\right)^{1/2} e^{\pm i\phi} \sin\theta$
- $Y_{20}(\theta, \phi) = \left(\frac{5}{16\pi}\right)^{1/2} \left(3\cos^2\theta - 1\right)$
- $Y_{2\pm1}(\theta, \phi) = \mp\left(\frac{15}{8\pi}\right)^{1/2} e^{\pm i\phi} \sin\theta$
- $Y_{2\pm2}(\theta, \phi) = \left(\frac{15}{32\pi}\right)^{1/2} e^{\pm 2i\phi} \sin^2\theta$

2 The wave function looks like this in spherical coordinates:

$\psi(r, \theta, \phi) = R_{nl}(r)\, Y_{lm}(\theta, \phi)$

And the Schrödinger equation looks like this:

$$-\frac{\hbar^2}{2m}\frac{1}{r}\frac{\partial^2}{\partial r^2} r\,\psi(r) + \frac{1}{2mr^2}\mathbf{L}^2\psi(r) + V(r)\psi(r) = E\psi(r)$$

You've already substituted the wave function into the Schrödinger equation and solved for $Y_{lm}(\theta, \phi)$. Now substitute the wave function into the Schrödinger equation and get an equation for $R_{nl}(r)$. **The answer is**

$$-\frac{\hbar^2}{2m}\frac{\partial^2}{\partial r^2}(r\,R_{nl}(r)) + \left(V(r) + \frac{l(l+1)\hbar^2}{2mr^2}\right)(r\,R_{nl}(r)) = E(r\,R_{nl}(r))$$

The Schrödinger equation looks like this:

$$-\frac{\hbar^2}{2m}\frac{1}{r}\frac{\partial^2}{\partial r^2} r\psi(r) + \frac{1}{2mr^2}\mathbf{L}^2\psi(r) + V(r)\psi(r) = E\psi(r)$$

And the wave function looks like this in general:

$\psi(r, \theta, \phi) = R_{nl}(r)\, Y_{lm}(\theta, \phi)$

Put the wave function into the Schrödinger equation, which gives you the following:

$$-\hbar^2\frac{r}{R_{nl}(r)}\frac{\partial^2}{\partial r^2}\left(r\,R_{nl}(r)\right) + 2mr^2\left(V(r) - E\right) + \frac{\mathbf{L}^2 Y_{lm}(\theta, \phi)}{Y_{lm}(\theta, \phi)} = 0$$

The spherical harmonics are eigenfunctions of $\mathbf{L}^2$ with eigenvalue $l(l+1)\hbar^2$, so

$\mathbf{L}^2 Y_{lm}(\theta, \phi) = l(l+1)\hbar^2 Y_{lm}(\theta, \phi)$

Therefore, the last term in the Schrödinger equation is simply $l(l+1)\hbar^2$. That means that you get

$$-\hbar^2\frac{r}{R_{nl}(r)}\frac{\partial^2}{\partial r^2}\left(r\,R_{nl}(r)\right) + 2mr^2\left(V(r) - E\right) + l(l+1)\hbar^2 = 0$$

Rewrite this as

$$-\frac{\hbar^2}{2m}\frac{d^2}{dr^2}\left(r\,R_{nl}(r)\right)+\left(V(r)+\frac{l(l+1)\hbar^2}{2mr^2}\right)\left(r\,R_{nl}(r)\right)=E\left(r\,R_{nl}(r)\right)$$

This is the equation you use to determine the radial part of the wave function, $R_{nl}(r)$.

3 The spherical Bessel functions are given by

$$j_l(\rho)=(-\rho)^l\left(\frac{1}{\rho}\frac{d}{d\rho}\right)^l\frac{\sin\rho}{\rho}$$

Find $j_0(\rho)$, $j_1(\rho)$, and $j_2(\rho)$. **Here are the answers:**

- $j_0(\rho)=\frac{\sin\rho}{\rho}$
- $j_1(\rho)=\frac{\sin\rho}{\rho^2}-\frac{\cos\rho}{\rho}$
- $j_2(\rho)=\frac{3\sin\rho}{\rho^3}-\frac{3\cos\rho}{\rho^2}-\frac{\sin\rho}{\rho}$

The spherical Bessel functions are given by

$$j_l(\rho)=(-\rho)^l\left(\frac{1}{\rho}\frac{d}{d\rho}\right)^l\frac{\sin\rho}{\rho}$$

Simply find $l=0$, $l=1$, and $l=2$.

4 The spherical Neumann functions are given by

$$n_l(\rho)=-(-\rho)^l\left(\frac{1}{\rho}\frac{d}{d\rho}\right)^l\frac{\cos\rho}{\rho}$$

Find $n_0(\rho)$, $n_1(\rho)$, and $n_2(\rho)$. **The answers are**

- $n_0(\rho)=-\frac{\cos\rho}{\rho}$
- $n_1(\rho)=-\frac{\cos\rho}{\rho^2}-\frac{\sin\rho}{\rho}$
- $n_2(\rho)=-\frac{3\cos\rho}{\rho^3}-\frac{3\sin\rho}{\rho^2}+\frac{\cos\rho}{\rho}$

The spherical Neumann functions are given by

$$n_l(\rho)=-(-\rho)^l\left(\frac{1}{\rho}\frac{d}{d\rho}\right)^l\frac{\cos\rho}{\rho}$$

To get the answers, solve the equations for $l=0$, $l=1$, and $l=2$.

5 Using the case where ρ is small, show that B must equal 0 in the radial part of the wave function solution for $0 < r < a$

$$R_l(\rho) = A_l j_l(\rho) + B_l n_l(\rho)$$

giving you this wave function:

$$\psi_{\text{inside}}(r, \theta, \phi) = A_l j_l(\rho_{\text{inside}})\, Y_{lm}(\theta, \phi)$$

The answer is $\psi_{\text{inside}}(r,\ ,\ ) = A_l j_l(\ _{\text{inside}})\, Y_{lm}(\ ,\ )$, where $\rho_{\text{inside}} = r\left(2m\left(E+V_0\right)\right)^{1/2}/\hbar$ and $Y_{lm}(\ ,\ )$ are the spherical harmonics.

You can apply the same constraint here that you use for a free particle — that the wave function must be finite everywhere. For small ρ, the Bessel functions look like this:

$$j_l(\rho) \approx \frac{2^l l! \rho^l}{\left(2^l+1\right)!}$$

And for small ρ, the Neumann functions reduce to

$$n_l(\rho) \approx \frac{-(2l-1)!\rho^{-l-1}}{2^l l!}$$

So the Neumann functions diverge for small ρ, which makes them unacceptable for wave functions here. That means that the radial part of the wave function is just made up of spherical Bessel functions, where A_l is a constant:

$$R_l(\rho) = A_l j_l(\rho)$$

The whole wave function inside the square well, $\psi_{\text{inside}}(r, \theta, \phi)$, is a product of radial and angular parts, and it looks like this:

$$\psi_{\text{inside}}(r, \theta, \phi) = A_l j_l(\rho_{\text{inside}})\, Y_{lm}(\theta, \phi)$$

where $\rho_{\text{inside}} = r\left(2m\left(E+V_0\right)\right)^{1/2}/\hbar$ and $Y_{lm}(\theta, \phi)$ are the spherical harmonics.

6 What's the wave function look like for $r > a$? Use boundary conditions (continuity of the wave function and its first derivative) to set up equations to find the wave function's normalization constants. **The answers are**

- **$_{\text{outside}}(r,\ ,\ ) = B_l(j_l(\ _{\text{outside}}) + n_l(\ _{\text{outside}}))\, Y_{lm}(\ ,\ )$**
- **$_{\text{inside}}(a,\ ,\ ) = \ _{\text{outside}}(a,\ ,\ )$**
- **$\frac{d}{dr}\psi_{\text{inside}}(r, \theta, \phi)\Big|_{r=a} = \frac{d}{dr}\psi_{\text{outside}}(r, \theta, \phi)\Big|_{r=a}$**

Outside the spherical well, in the region $r > a$, the particle is just like a free particle, so here's what the radial equation looks like:

$$-\frac{\hbar^2}{2m}\frac{\partial^2}{\partial r^2}\left(r\, R_{nl}(r)\right) + \left(V(r) + \frac{l(l+1)\hbar^2}{2mr^2}\right)\left(r\, R_{nl}(r)\right) = E\left(r\, R_{nl}(r)\right)$$

You solve this equation earlier in the chapter (see the section "Dealing Freely with Free Particles in Spherical Coordinates") — you make the change of variable $\rho = kr$, where $k = (2mE)^{1/2}/\hbar$, so that $R_{nl}(r)$ becomes $R_l(kr) = R_l(\rho)$. Using this substitution, you get

$$\frac{d^2R_l(\rho)}{d\rho^2}+\frac{2}{\rho}\frac{dR_l(\rho)}{d\rho}+\left(1-\frac{l(l+1)}{\rho^2}\right)R_l(\rho)=0$$

And the solution is a combination of spherical Bessel functions and spherical Neumann functions:

$$R_l(\rho) = B_l(j_l(\rho) + n_l(\rho))$$

where B_l is a constant.

The radial solution outside the bounds of the square well looks like this, where $\rho_{\text{outside}} = r(2mE)^{1/2}/\hbar$:

$$\psi_{\text{outside}}(r, \theta, \phi) = B_l(j_l(\rho_{\text{outside}}) + n_l(\rho_{\text{outside}}))\, Y_{lm}(\theta, \phi)$$

You know that the wave function inside the bounds of the square well is

$$\psi_{\text{inside}}(r, \theta, \phi) = A_l j_l(\rho_{\text{inside}})\, Y_{lm}(\theta, \phi)$$

You find A_l and B_l through continuity constraints. At the inside/outside boundary, where $r = a$, the wave function and its first derivative must be continuous. So to determine A_l and B_l, you have to solve these two equations:

- $\psi_{\text{inside}}(a, \theta, \phi) = \psi_{\text{outside}}(a, \theta, \phi)$
- $\frac{d}{dr}\psi_{\text{inside}}(r, \theta, \phi)\Big|_{r=a} = \frac{d}{dr}\psi_{\text{outside}}(r, \theta, \phi)\Big|_{r=a}$

7 The radial part of the wave function for isotropic harmonic oscillators relies on the generalized Laguerre polynomials:

$$L_a^b(r)=\frac{r^{-b}e^r}{a!}\frac{d^a}{dr^a}\left(e^{-r}r^{a+b}\right)$$

Find $L_0{}^b(r)$, $L_1{}^b(r)$, $L_2{}^b(r)$, and $L_3{}^b(r)$. **The answers are**

- $\mathbf{L_0{}^b(r) = 1}$
- $\mathbf{L_1{}^b(r) = -r + b + 1}$
- $\mathbf{L_2^b(r)=-\frac{r^2}{2}-(b+2)r+\frac{(b+2)(b+1)}{2}}$
- $\mathbf{L_3^b(r)=-\frac{r^3}{6}-\frac{(b+3)r^2}{2}-\frac{(b+2)(b+3)r}{2}+\frac{(b+1)(b+2)(b+3)}{6}}$

The generalized Laguerre polynomials are given by

$$L_a^b(r)=\frac{r^{-b}e^r}{a!}\frac{d^a}{dr^a}\left(e^{-r}r^{a+b}\right)$$

Find $a = 0, 1, 2,$ and 3 to get the answers.

8 Find the full expressions for $\psi_{110}(r, \theta, \phi)$, $\psi_{11\pm1}(r, \theta, \phi)$, and $\psi_{200}(r, \theta, \phi)$ for a particle of mass m in an isotopic harmonic oscillator. **Here are the answers:**

- $$\psi_{110}(r,\theta,\phi)=\frac{\left(\frac{8}{3}\right)^{1/2}}{\pi^{1/4}}\left(\frac{m\omega}{\hbar}\right)^{5/4} r\exp\left(\frac{-m\omega r^2}{2\hbar}\right)\left(\frac{3}{4\pi}\right)^{1/2}\cos\theta$$
- $$\psi_{11\pm1}(r,\theta,\phi)=\mp\frac{\left(\frac{8}{3}\right)^{1/2}}{\pi^{1/4}}\left(\frac{m\omega}{\hbar}\right)^{5/4} r\exp\left(\frac{-m\omega r^2}{2\hbar}\right)\left(\frac{3}{8\pi}\right)^{1/2} e^{\pm i\phi}\sin\theta$$
- $$\psi_{200}(r,\theta,\phi)=\mp\frac{\left(\frac{8}{3}\right)^{1/2}}{\pi^{1/4}}\left(\frac{m\omega}{\hbar}\right)^{3/4} r\exp\left(\frac{3}{2}-\frac{m\omega r^2}{\hbar}\right)\exp\left(\frac{-m\omega r^2}{2\hbar}\right)\frac{1}{(4\pi)^{1/2}}$$

The general form for the wave function is

$$\psi_{nlm}(r,\theta,\phi)=R_{nl}(r)\,Y_{lm}(\theta,\phi)$$

For an isotropic harmonic oscillator, the radial part of the wave function looks like this:

$$R_{nl}(r)=C_{nl}r^{l}\exp\left(\frac{-m\omega r^2}{2\hbar}\right)L^{(l+1/2)}_{(n-l)/2}\left(\frac{m\omega r^2}{\hbar}\right)$$

where $\exp(x) = e^x$ and $C_{nl}=\frac{\left(2^{n+l+2}(m\omega/\hbar)^{l+3/2}/\pi^{1/2}\right)^{1/2}\left[(n-l)/2\right]!\left[(n+l)/2\right]!}{(n+l+1)!^{1/2}}$ and the $L_a^{\,b}(r)$ functions are the generalized Laguerre polynomials:

$$L_a^b(r)=\frac{r^{-b}e^{r}}{a!}\frac{d^a}{dr^a}\left(e^{-r}r^{a+b}\right)$$

Substitute the radial part into the wave function, which gives you the following:

- $$\psi_{000}(r,\theta,\phi)=\frac{2}{\pi^{1/4}}\left(\frac{m\omega}{\hbar}\right)^{3/4}\exp\left(\frac{-m\omega r^2}{2\hbar}\right)Y_{00}(\theta,\phi)$$
- $$\psi_{1m}(r,\theta,\phi)=\frac{\left(\frac{8}{3}\right)^{1/2}}{\pi^{1/4}}\left(\frac{m\omega}{\hbar}\right)^{5/4} r\exp\left(\frac{-m\omega r^2}{2\hbar}\right)Y_{lm}(\theta,\phi)$$
- $$\psi_{200}(r,\theta,\phi)=\frac{\left(\frac{8}{3}\right)^{1/2}}{\pi^{1/4}}\left(\frac{m\omega}{\hbar}\right)^{3/4}\left(\frac{3}{2}-\frac{m\omega r^2}{\hbar}\right)\exp\left(\frac{-m\omega r^2}{2\hbar}\right)Y_{00}(\theta,\phi)$$

And here are the spherical harmonics you need:

- $$Y_{00}(\theta,\phi)=\frac{1}{(4\pi)^{1/2}}$$
- $Y_{10}(\theta, \phi) = (3/4\pi)^{1/2}\cos\theta$
- $$Y_{1\pm1}(\theta,\phi)=\mp\left(\frac{3}{8\pi}\right)^{1/2}e^{\pm i\phi}\sin\theta$$

Putting the spherical harmonics into the wave function finally gives you

- $\psi_{110}(r, \theta, \phi) = \frac{\left(\frac{8}{3}\right)^{1/2}}{\pi^{1/4}} \left(\frac{m\omega}{\hbar}\right)^{5/4} r \exp\left(\frac{-m\omega r^2}{2\hbar}\right) \left(\frac{3}{4\pi}\right)^{1/2} \cos\theta$
- $\psi_{11\pm1}(r, \theta, \phi) = \mp \frac{\left(\frac{8}{3}\right)^{1/2}}{\pi^{1/4}} \left(\frac{m\omega}{\hbar}\right)^{5/4} r \exp\left(\frac{-m\omega r^2}{2\hbar}\right) \left(\frac{3}{8\pi}\right)^{1/2} e^{\pm i\phi} \sin\theta$
- $\psi_{200}(r, \theta, \phi) = \frac{\left(\frac{8}{3}\right)^{1/2}}{\pi^{1/4}} \left(\frac{m\omega}{\hbar}\right)^{3/4} \left(\frac{3}{2} - \frac{m\omega r^2}{\hbar}\right) \exp\left(\frac{-m\omega r^2}{2\hbar}\right) \frac{1}{(4\pi)^{1/2}}$

Chapter 8

Getting to Know Hydrogen Atoms

In This Chapter

- Understanding the Schrödinger equation for hydrogen
- Using center-of-mass coordinates
- Breaking up hydrogen's Schrödinger equation into solvable parts
- Working with hydrogen's radial wave functions
- Finding energy levels and energy degeneracy

The hydrogen atom is one of the successes of quantum physics. When you get to multi-electron atoms, the situation becomes much harder to handle because all the electrons can interact with each other as well as with the nucleus. But the hydrogen atom presents you with a relatively simple case that you can make a lot of progress with, solving for wave functions and energy levels.

This chapter considers problems on the hydrogen atom, the most basic of all the atoms, with only an electron and a proton. Here you see how to create the Schrödinger equation for the hydrogen atom, solve it for the wave functions, and determine the energy levels.

Eyeing How the Schrödinger Equation Appears for Hydrogen

In a hydrogen atom, an electron circles a proton, and the whole thing is held together by electric forces. Figure 8-1 shows a hydrogen atom.

To get anywhere with the hydrogen atom system in quantum physics, you have to construct the Schrödinger equation, which you can then solve to get the wave functions. When dealing with hydrogen atom systems, the Schrödinger equation looks like this:

$$-\frac{\hbar^2}{2m_p}\nabla_p^2\psi(\mathbf{r}_e,\mathbf{r}_p)-\frac{\hbar^2}{2m_e}\nabla_e^2\psi(\mathbf{r}_e,\mathbf{r}_p)-\frac{e^2}{|\mathbf{r}_e-\mathbf{r}_p|}\psi(\mathbf{r}_e,\mathbf{r}_p)=E\psi(\mathbf{r}_e,\mathbf{r}_p)$$

The following example shows you how different energy levels can alter the Schrödinger equation, and the practice problems allow you to work with the electron's kinetic energy and the atom's potential energy.

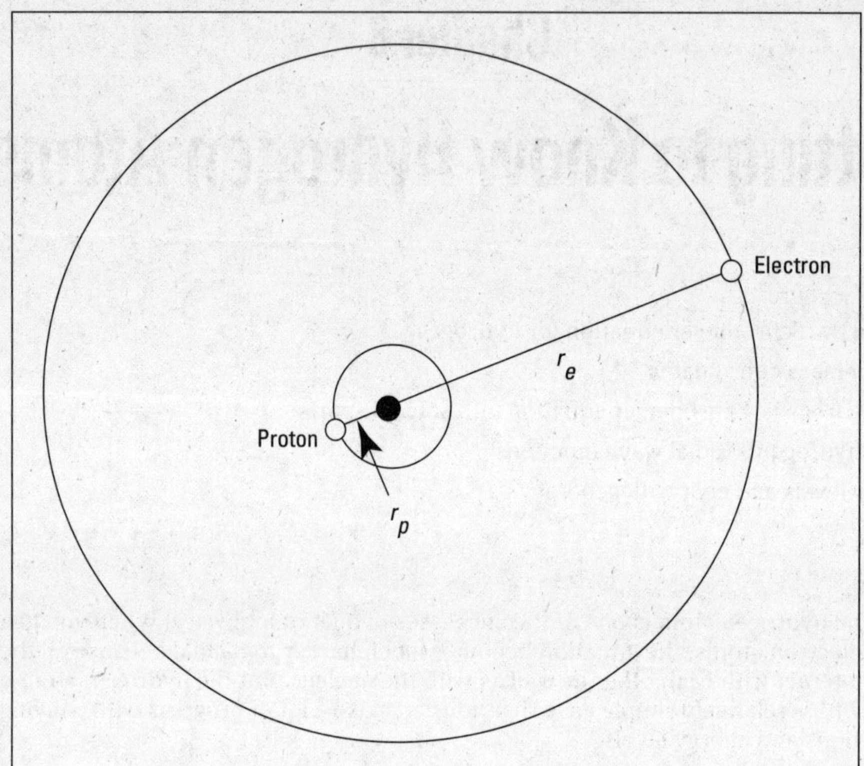

Figure 8-1: The hydrogen atom.

Q. What does the kinetic energy of the proton look like in the Schrödinger equation?

A. $-\frac{\hbar^2}{2m_p}\nabla_p^2$, **where** $\nabla_p^2 = \frac{\partial^2}{\partial x_p^2} + \frac{\partial^2}{\partial y_p^2} + \frac{\partial^2}{\partial z_p^2}$.

Start with the Schrödinger equation:

$$-\frac{\hbar^2}{2m_p}\nabla_p^2\psi(\mathbf{r}_e, \mathbf{r}_p) - \frac{\hbar^2}{2m_e}\nabla_e^2\psi(\mathbf{r}_e, \mathbf{r}_p) - \frac{e^2}{|\mathbf{r}_e - \mathbf{r}_p|}\psi(\mathbf{r}_e, \mathbf{r}_p) = E\psi(\mathbf{r}_e, \mathbf{r}_p)$$

Now just separate out the term that has to do with the kinetic energy of the proton — the first term:

$$-\frac{\hbar^2}{2m_p}\nabla_p^2, \text{ where } \nabla_p^2 = \frac{\partial^2}{\partial x_p^2} + \frac{\partial^2}{\partial y_p^2} + \frac{\partial^2}{\partial z_p^2}$$

Here, x_p is the proton's x position, y_p is the proton's y position, and so on.

1. What does the electron's kinetic energy term look like in the hydrogen atom's Schrödinger equation?

Solve It

2. What does the potential energy term look like in the hydrogen atom's Schrödinger equation?

Solve It

Switching to Center-of-Mass Coordinates to Make the Hydrogen Atom Solvable

Solving the hydrogen Schrödinger equation from the preceding section isn't easy because it contains terms like $|\mathbf{r}_e - \mathbf{r}_p|$. Solving it would be a lot easier if you could assume that the proton was stationary and that $r_p = 0$, which would give you

$$-\frac{\hbar^2}{2m_e}\nabla_e^2\psi(\mathbf{r}_e) - \frac{e^2}{|\mathbf{r}_e|}\psi(\mathbf{r}_e) = E\psi(\mathbf{r}_e)$$

However, that equation is inaccurate because the proton isn't stationary. It rotates around the electron just as the electron rotates around it.

How can you convert the full Schrödinger equation into something manageable like this without sacrificing accuracy? You can use center-of-mass coordinates. *Center-of-mass coordinates* have the center of mass at the origin, and these coordinates are a good choice when you have two moving particles.

Q. Relate the Laplacian operators for the electron's position, ∇_e^2, and the proton's position, ∇_p^2, to the same operators using center-of-mass coordinates, ∇_R^2 and ∇_r^2, where

$$\mathbf{R} = \frac{m_e\mathbf{r}_e + m_p\mathbf{r}_p}{m_e + m_p}$$

and the vector between the electron and proton is $\mathbf{r} = \mathbf{r}_e - \mathbf{r}_p$.

A. $\frac{1}{m_e}\nabla_e^2 + \frac{1}{m_p}\nabla_p^2 = \frac{1}{M}\nabla_R^2 + \frac{1}{m}\nabla_r^2$**, where $M = m_e + m_p$ is the total mass and $m = \frac{m_e m_p}{m_e + m_p}$ is the reduced mass.**

The center of mass of the proton/electron system is at the following location:

$$\mathbf{R} = \frac{m_e\mathbf{r}_e + m_p\mathbf{r}_p}{m_e + m_p}$$

And the vector between the electron and proton is $\mathbf{r} = \mathbf{r}_e - \mathbf{r}_p$. Using $\mathbf{R}$ and $\mathbf{r}$ in the Schrödinger equation instead of $\mathbf{r}_e$ and $\mathbf{r}_p$ is going to make that equation easier to solve.

The Laplacian for $\mathbf{R}$ in an x, y, z coordinate system is

$$\nabla_R^2 = \frac{\partial^2}{\partial X^2} + \frac{\partial^2}{\partial Y^2} + \frac{\partial^2}{\partial Z^2}$$

And the Laplacian for $\mathbf{r}$ is

$$\nabla_r^2 = \frac{\partial^2}{\partial x^2} + \frac{\partial^2}{\partial y^2} + \frac{\partial^2}{\partial z^2}$$

Rewrite ∇_e^2 and ∇_p^2 in terms of ∇_R^2 and ∇_r^2. Doing so gives you

$$\frac{1}{m_e}\nabla_e^2 + \frac{1}{m_p}\nabla_p^2 = \frac{1}{M}\nabla_R^2 + \frac{1}{m}\nabla_r^2$$

where $M = m_e + m_p$ is the total mass and $m = \frac{m_e m_p}{m_e + m_p}$ is called the *reduced mass*, which is the effective mass of the electron-proton system.

3. What is the Schrödinger equation in terms of $\nabla_{\mathbf{R}}^2$ and $\nabla_{\boldsymbol{r}}^2$?

Solve It

4. Separate the Schrödinger equation

$$-\frac{\hbar^2}{2M}\nabla_{\mathbf{R}}^2\psi(\mathbf{R},\boldsymbol{r})-\frac{\hbar^2}{2m}\nabla_{\boldsymbol{r}}^2\psi(\mathbf{R},\boldsymbol{r})-\frac{e^2}{|\boldsymbol{r}|}\psi(\mathbf{R},\boldsymbol{r})=E\psi(\mathbf{R},\boldsymbol{r})$$

into two equations, one for $\boldsymbol{r}$ and one for $\mathbf{R}$, where $\mathbf{R}=\frac{m_e\boldsymbol{r}_e+m_p\boldsymbol{r}_p}{m_e+m_p}$ and $\boldsymbol{r}=\boldsymbol{r}_e-\boldsymbol{r}_p$. (***Hint:*** Use the substitution $\psi(\mathbf{R},\boldsymbol{r})=\psi(\mathbf{R})\psi(\boldsymbol{r})$).

Solve It

Doing the Splits: Solving the Dual Schrödinger Equation

When you encounter a complicated Schrödinger equation, splitting it up is the best way to handle it. You can divide the Schrödinger equation for the hydrogen atom into two equations, so you can split it:

$$-\frac{\hbar^2}{2M}\nabla_R^2\psi(\mathbf{R}) = E_R\psi(\mathbf{R})$$

Similarly, you get this for $\psi(\boldsymbol{r})$:

$$-\frac{\hbar^2}{2m}\nabla_r^2\psi(r) - \frac{e^2}{|\boldsymbol{r}|}\psi(r) = E_r\psi(r)$$

where $\boldsymbol{r} = \boldsymbol{r}_e - \boldsymbol{r}_p$ is the vector between the electron and proton.

In this way, you have two equations: one for the center of mass and one for the vector between the electron and proton.

Now you get to solve this dual equation. The example starts with the equation for $\psi(\mathbf{R})$.

Q. Solve the following Schrödinger equation for $\psi(\mathbf{R})$:

$$-\frac{\hbar^2}{2M}\nabla_R^2\psi(\mathbf{R}) = E_R\psi(\mathbf{R})$$

A. $\psi(\mathbf{R}) = \frac{e^{-i\boldsymbol{k}\cdot\boldsymbol{r}}}{(2\pi)^{3/2}}$

The solution to this differential Schrödinger equation is

$\psi(\mathbf{R}) = Ce^{-i\boldsymbol{k}\cdot\boldsymbol{r}}$

where C is a constant and $\boldsymbol{k}$ is the wave vector, where $|\boldsymbol{k}| = \left(\frac{2ME_R}{\hbar^2}\right)^{1/2}$.

Optionally, you can find C by insisting that $\psi(\mathbf{R})$ be normalized, as all wave functions must, which means that

$$1 = \int_0^\infty \psi(\mathbf{R})\psi^*(\mathbf{R})d^3\mathbf{R}$$

In other words, $C = \frac{1}{(2\pi)^{3/2}}$. Plug C into $\psi(\mathbf{R}) = Ce^{-i\boldsymbol{k}\cdot\boldsymbol{r}}$ to get the final answer:

$$\psi(\mathbf{R}) = \frac{e^{-i\boldsymbol{k}\cdot\boldsymbol{r}}}{(2\pi)^{3/2}}$$

5. The Schrödinger equation for $\psi(\boldsymbol{r})$ is

$$-\frac{\hbar^2}{2m}\nabla_r^2\psi(\boldsymbol{r})-\frac{e^2}{|\boldsymbol{r}|}\psi(\boldsymbol{r})=E_r\psi(\boldsymbol{r})$$

where $\boldsymbol{r} = \boldsymbol{r}_e - \boldsymbol{r}_p$. You can break the solution, $\psi(\boldsymbol{r})$, into a radial and an angular part (see Chapter 7 on spherical coordinates for more on this process):

$\psi(\boldsymbol{r}) = \mathbf{R}_{nl}(\boldsymbol{r})\, Y_{lm}(\theta, \phi)$

The angular part of $\psi(\boldsymbol{r})$ is made up of spherical harmonics, $Y_{lm}(\theta, \phi)$ (refer to Chapter 7). Now you have to solve for the radial part, $R_{nl}(\boldsymbol{r})$.

Here's what the Schrödinger equation becomes for the radial part:

$$-\frac{\hbar^2}{2m}\frac{d^2}{dr^2}r\,R_{nl}(r)+l(l+1)\frac{\hbar^2}{2mr^2}r\,R_{nl}(r)-\frac{e^2}{r}r\,R_{nl}(r)=E_r\,r\,R_{nl}(r)$$

where $r = |\boldsymbol{r}|$. Solve this equation for small r up to an arbitrary constant.

Solve It

6. The radial Schrödinger equation becomes the following for the radial part of $\psi(\boldsymbol{r})$:

$$-\frac{\hbar^2}{2m}\frac{d^2}{dr^2}r\,R_{nl}(r)+l(l+1)\frac{\hbar^2}{2mr^2}r\,R_{nl}(r)-\frac{e^2}{r}r\,R_{nl}(r)=E_r\,r\,R_{nl}(r)$$

where $\boldsymbol{r} = \boldsymbol{r}_e - \boldsymbol{r}_p$ and $r = |\boldsymbol{r}|$. Solve this equation for large R up to an arbitrary constant.

Solve It

Solving the Radial Schrödinger Equation for ψ(r)

After you divide the Schrödinger equation into radial and angular parts (see the preceding section), you need to know how to solve for the radial part of the Schrödinger equation for a hydrogen atom, $R_{nl}(r) \approx Ar^l$ for small r and $R_{nl}(r) \approx Ae^{-\lambda r}$ for large r (see problems 5 and 6 for the calculations). Putting these together gives you the solution to the radial Schrödinger equation:

$$R_{nl}(r) = r^l f(r) e^{-\lambda r}$$

where f(r) is some as-yet undetermined function of r. You can determine f(r) by substituting this form for $R_{nl}(r)$ into the radial Schrödinger equation and seeing what form for f(r) solves that equation:

$$-\frac{\hbar^2}{2m}\frac{d^2}{dr^2} r\, R_{nl}(r) + l(l+1)\frac{\hbar^2}{2mr^2} r\, R_{nl}(r) - \frac{e^2}{r} r\, R_{nl}(r) = E_r\, r\, R_{nl}(r)$$

Plugging in the value of $R_{nl}(r)$ gives you

$$\frac{d^2}{dr^2}f(r) + 2\left(\frac{l(l+1)}{r} - \lambda\right)\frac{df(r)}{dr} + 2\left(\frac{\frac{me^2}{\hbar^2} - \lambda(l+1)}{r}\right)f(r) = 0$$

The usual way to solve this is to use a series expansion for f(r) like this, where k is the series index value and a_k represents coefficients:

$$f(r) = \sum_{k=0}^{\infty} a_k r^k$$

The following example looks into solving for $R_{nl}(r)$.

Q. The radial part of the wave function ψ(r) looks like $R_{nl}(r) = r^l f(r) e^{-\lambda r}$. Use this form for f($r$):

$$f(r) = \sum_{k=0}^{\infty} a_k r^k$$

which gives you the following form of the Schrödinger equation:

$$\frac{d^2}{dr^2}f(r) + 2\left(\frac{l(l+1)}{r} - \lambda\right)\frac{df(r)}{dr} + 2\left(\frac{\frac{me^2}{\hbar^2} - \lambda(l+1)}{r}\right)f(r) = 0$$

Given that $\psi_{nlm}(r, \theta, \phi) = R_{nl}(r)\, Y_{lm}(\theta, \phi)$, solve for $\psi_{100}(r, \theta, \phi)$.

A. $\boldsymbol{\psi_{100}(r, \theta, \phi) = \frac{2}{r_0^{3/2}} e^{-r/r_0} Y_{00}(\theta, \phi)}$, **where** $\boldsymbol{r_0 = \frac{\hbar^2}{me^2}}$.

Substitute $R_{10}(r)$ into the Schrödinger equation, which gives you the following:

$$\sum_{k=0}^{\infty}\left[k(k+2l+1)a_k r^{k-2}+2\left(\frac{me^2}{\hbar^2}-\lambda(k+l+1)\right)a_k r^{k-1}\right]=0$$

Change the index of the second term from k to $k-1$:

$$\sum_{k=0}^{\infty}\left[k(k+2l+1)a_k r^{k-2}+2\left(\frac{me^2}{\hbar^2}-\lambda(k+l)\right)a_{k-1}r^{k-2}\right]=0$$

Because each term in this series has to be zero to have each power of r be zero (because the whole series equals zero), you get the following:

$$k(k+2l+1)a_k r^{k-2}=2\left(\lambda(k+l)-\frac{me^2}{\hbar^2}\right)a_{k-1}r^{k-2}$$

Divide by r^{k-2}, which gives you

$$k(k+2l+1)a_k=2\left(\lambda(k+l)-\frac{me^2}{\hbar^2}\right)a_{k-1}$$

Now take a look at the ratio of a_k/a_{k-1}:

$$\frac{a_k}{a_{k-1}}=\frac{2\left(\lambda(k+l)-\frac{me^2}{\hbar^2}\right)}{k(k+2l+1)}$$

This resembles the expansion for e^{2x}, which is

$$e^x=\sum_{k=0}^{\infty}\frac{(2x)^k}{k!}$$

Therefore, you may suspect that $f(r)$ is an exponential.

The radial wave function, $R_{nl}(r)$, looks like this:

$$R_{nl}(r) = r^l f(r)e^{-\lambda r}$$

where $\lambda=\frac{(-2mE)^{1/2}}{\hbar}$.

Trying a form of $f(r)$ like $f(r) = e^{2\lambda r}$ gives you the following:

$$R_{nl}(r) = r^l e^{2\lambda r}e^{-\lambda r}$$

which equals

$$R_{nl}(r) = r^l e^{\lambda r}$$

This solution has a problem: It goes to infinity as r goes to infinity. So now try a solution for $f(r)$ that looks like this (note that the summation is now to N, not ∞):

$$f(r)=\sum_{k=0}^{N}a_k r^k$$

For this series to terminate, a_{N+1}, a_{N+2}, a_{N+3}, and so on must all be zero. Here's the recurrence relation for the coefficients a_k:

$$k(k+2l+1)a_k = 2\left(\lambda(k+l) - \frac{me^2}{\hbar^2}\right)a_{k-1}$$

For a_{N+1} to be zero, the factor multiplying a_{k-1} must be zero for $k = N + 1$, which means that

$$2\left(\lambda(k+l) - \frac{me^2}{\hbar^2}\right) = 0 \quad \text{for } k = N+1$$

Substitute in $k = N + 1$, which gives you the following:

$$2\left(\lambda(N+l+1) - \frac{me^2}{\hbar^2}\right) = 0$$

Divide by 2 to get

$$\lambda(N+l+1) - \frac{me^2}{\hbar^2} = 0$$

Substitute $N + l + 1 \rightarrow n$, where n is the *principal quantum number.* Doing so gives you

$$n\lambda - \frac{me^2}{\hbar^2} = 0 \quad n = 1, 2, 3, \ldots$$

This is the quantization condition that must be met. So here's the form you have for $R_{nl}(r)$:

$$R_{nl}(r) = A_{nl} r^l e^{-\lambda r} \sum_{k=0}^{N} a_k r^k$$

You find A_{nl} by normalizing $R_{nl}(r)$. You normalize $R_{10}(r)$ like this:

$$1 = \int_0^{+\infty} r^2 \left|R_{10}(r)\right|^2 dr$$

which gives you

$$R_{10}(r) = \frac{2}{r_0^{3/2}} e^{-r/r_0}$$

where $r_0 = \frac{\hbar^2}{me^2}$.

You know that $\psi_{nlm}(r, \theta, \phi)$ equals

$$\psi_{nlm}(r, \theta, \phi) = R_{nl}(r)\, Y_{lm}(\theta, \psi)$$

So $\psi_{100}(r, \theta, \phi)$ becomes the following:

$$\psi_{100}(r, \theta, \phi) = \frac{2}{r_0^{3/2}} e^{-r/r_0} Y_{00}(\theta, \phi)$$

7. Given that the hydrogen atom wave functions look like this:

$$\psi_{nlm}(r,\theta,\phi)=\frac{\left(\frac{2}{nr_0}\right)^{3/2}\left((n-l-1)!\right)^{1/2}}{\left(2n(n+1)!\right)^{1/2}}e^{-r/(nr_0)}\left(\frac{2r}{nr_0}\right)^{l}L_{n-l-1}^{2l+1}\left(\frac{2r}{nr_0}\right)Y_{lm}(\theta,\phi)$$

where $r_0=\frac{\hbar^2}{me^2}$ and where $L_{n-l-1}^{2l+1}(2r/(nr_0))$ is a generalized Laguerre polynomial:

- $L_0^b(r) = 1$
- $L_1^b(r) = -r + b + 1$
- $L_2^b(r)=-\frac{r^2}{6}-(b+2)r+\frac{(b+2)(b+1)}{2}$
- $L_3^b(r)=-\frac{r^3}{6}-\frac{(b+3)r^2}{2}-\frac{(b+2)(b+3)r}{2}+\frac{(b+1)(b+2)(b+3)}{6}$

find $\psi_{200}(r,\theta,\phi)$.

Solve It

8. Given that the hydrogen atom wave functions look like this:

$$\psi_{nlm}(r,\theta,\phi)=\frac{\left(\frac{2}{nr_0}\right)^{3/2}\left((n-l-1)!\right)^{1/2}}{\left(2n(n+1)!\right)^{1/2}}e^{-r/(nr_0)}\left(\frac{2r}{nr_0}\right)^{l}\mathrm{L}_{n-l-1}^{2l+1}\left(\frac{2r}{nr_0}\right)\mathrm{Y}_{lm}(\theta,\phi)$$

where $r_0=\frac{\hbar^2}{me^2}$ and where $\mathrm{L}_{n-l-1}^{2l+1}\left(2r/(nr_0)\right)$ is a generalized Laguerre polynomial:

- $\mathrm{L}_0^b(r)=1$
- $\mathrm{L}_1^b(r)=-r+b+1$
- $\mathrm{L}_2^b(r)=-\frac{r^2}{6}-(b+2)r+\frac{(b+2)(b+1)}{2}$
- $\mathrm{L}_3^b(r)=-\frac{r^3}{6}-\frac{(b+3)r^2}{2}-\frac{(b+2)(b+3)r}{2}+\frac{(b+1)(b+2)(b+3)}{6}$

find $\psi_{300}(r,\theta,\phi)$.

Solve It

Juicing Up the Hydrogen Energy Levels

You can figure out the hydrogen energy levels using the wave functions. To make sure that $\psi(r)$ stays finite for the hydrogen atom, you need to have

$$n\lambda - \frac{me^2}{\hbar^2} = 0 \quad n = 1, 2, 3, \ldots$$

where $\lambda = \frac{(-2mE)^{1/2}}{\hbar}$.

This quantization condition (which I show you how to find in the example problem in the preceding section) actually constrains the possible values of energy that the hydrogen system can take. The following example and practice problems let you solve for those energy levels.

Q. Find a quantization condition in a single equation for the energy levels of hydrogen.

A. $$n\frac{(-2mE)^{1/2}}{\hbar} = \frac{me^2}{\hbar^2} \quad n = 1, 2, 3, \ldots$$

The quantization condition for $\psi(r)$ to remain finite as $r \rightarrow \infty$ is

$$n\lambda - \frac{me^2}{\hbar^2} = 0 \quad n = 1, 2, 3, \ldots$$

where $\lambda = \frac{(-2mE)^{1/2}}{\hbar}$.

Substitute λ into the first equation, which gives you the following:

$$n\frac{(-2mE)^{1/2}}{\hbar} - \frac{me^2}{\hbar^2} = 0 \quad n = 1, 2, 3, \ldots$$

Therefore

$$n\frac{(-2mE)^{1/2}}{\hbar} = \frac{me^2}{\hbar^2} \quad n = 1, 2, 3, \ldots$$

9. Solve for the energy levels of hydrogen in terms of the quantization number *n*.

Solve It

10. Find the first three energy levels of hydrogen numerically.

Solve It

Doubling Up on Energy Level Degeneracy

You can have *energy degeneracy* — that is, various states with the same energy — if you take into account *m,* the *z* component of the angular momentum. The energy of the hydrogen atom is dependent only on *n,* the principal quantum number:

$$E_n = -\frac{me^4}{2n^2\hbar^2} \quad n = 1, 2, 3, \ldots$$

That means the E is independent of *l* and *m,* the *z* component of the angular momentum. How many states, $\hbar^2$, have the same energy for a particular value of *n?* That's called the *angular momentum degeneracy* of hydrogen. The example problem calculates how degenerate each *n* level in hydrogen is. (For details on angular momentum, flip to Chapter 5.)

Q. Find the angular momentum degeneracy of the hydrogen atom in terms of *n.*

A. $\textbf{Degeneracy} = \sum_{l=0}^{n-1} (2l+1) = n^2$

For a specific value of *n, l* can range from zero to $n - 1$. And each *l* can have different values of *m* (from $-l$ to $+l$), so the total degeneracy is

$$\text{Degeneracy} = \sum_{l=0}^{n-1} (\text{Degeneracy in } m)$$

For a particular value of *l,* you can have *m* values of $-l, -l + 1, \ldots, 0, \ldots, l - 1, l$. So you can enter in $(2l + 1)$ for the degeneracy in *m:*

$$\text{Degeneracy} = \sum_{l=0}^{n-1} (2l+1)$$

This series is just n^2:

$$\text{Degeneracy} = \sum_{l=0}^{n-1} (2l+1) = n^2$$

So the angular momentum degeneracy of the energy levels of the hydrogen atom is n^2.

11. Verify that the angular momentum degeneracy of the $n = 1$ and $n = 2$ states is n^2.

Solve It

12. Find the degeneracy of the hydrogen atom when you consider spin in addition to angular momentum. (See Chapter 5 for info on spin.)

Solve It

Answers to Problems on Hydrogen Atoms

Here are the answers to the practice questions I present earlier in this chapter, along with the original questions and answer explanations.

1 What does the electron's kinetic energy term look like in the hydrogen atom's Schrödinger equation? **The answer is $-\frac{\hbar^2}{2m_e}\nabla_e^2$, where $\nabla_e^2 = \frac{\partial^2}{\partial x_e^2} + \frac{\partial^2}{\partial y_e^2} + \frac{\partial^2}{\partial z_e^2}$.**

The Schrödinger equation includes a term for the electron's kinetic energy:

$$-\frac{\hbar^2}{2m_e}\nabla_e^2$$

where $\nabla_e^2 = \frac{\partial^2}{\partial x_e^2} + \frac{\partial^2}{\partial y_e^2} + \frac{\partial^2}{\partial z_e^2}$. Here, x_e is the electron's x position, y_e is the electron's y position, and so on.

2 What does the potential energy term look like in the hydrogen atom's Schrödinger equation? **Here's the answer:**

$$\mathbf{V}(\boldsymbol{r}) = -\frac{e^2}{|\boldsymbol{r}_e - \boldsymbol{r}_p|}$$

The potential energy, V(**r**), is the third term in the Schrödinger equation:

$$-\frac{\hbar^2}{2m_p}\nabla_p^2\psi(\boldsymbol{r}_e, \boldsymbol{r}_p) - \frac{\hbar^2}{2m_e}\nabla_e^2\psi(\boldsymbol{r}_e, \boldsymbol{r}_p) + \mathrm{V}(\boldsymbol{r})\psi(\boldsymbol{r}_e, \boldsymbol{r}_p) = \mathrm{E}\psi(\boldsymbol{r}_e, \boldsymbol{r}_p)$$

where $\psi(\boldsymbol{r}_e, \boldsymbol{r}_p)$ is the electron and proton's wave function. The electrostatic potential energy for a central potential is given by

$$\mathrm{V}(\boldsymbol{r}) = -\frac{1}{4\pi\varepsilon_0}\frac{e^2}{|\boldsymbol{r}|}$$

where $\boldsymbol{r}$ is the radius vector separating the two charges. As is common in quantum mechanics, use CGS (centimeter-gram-second) units, where

$$1 = \frac{1}{4\pi\varepsilon_0}$$

Therefore, the potential due to the electron and proton charges in the hydrogen atom is

$$\mathrm{V}(\boldsymbol{r}) = -\frac{\mathrm{e}^2}{|\boldsymbol{r}|}$$

Because $\boldsymbol{r} = \boldsymbol{r}_e - \boldsymbol{r}_p$, this becomes

$$\mathrm{V}(\boldsymbol{r}) = -\frac{e^2}{|\boldsymbol{r}_e - \boldsymbol{r}_p|}$$

3 What is the Schrödinger equation in terms of ∇_R^2 and ∇_r^2?

The answer is $-\frac{\hbar^2}{2M}\nabla_R^2\psi(\mathbf{R}, \boldsymbol{r}) - \frac{\hbar^2}{2m}\nabla_r^2\psi(\mathbf{R}, \boldsymbol{r}) - \frac{e^2}{|\boldsymbol{r}|}\psi(\mathbf{R}, \boldsymbol{r}) = E\psi(\mathbf{R}, \boldsymbol{r})$

where $\mathbf{R} = \frac{m_e r_e + m_p r_p}{m_e + m_p}$, $\boldsymbol{r} = \boldsymbol{r}_e - \boldsymbol{r}_p$, $M = m_e + m_p$, **and** $m = \frac{m_e m_p}{m_e + m_p}$.

The example problem relates ∇_e^2 and ∇_p^2 to ∇_R^2 and ∇_r^2 this way:

$$\frac{1}{m_e}\nabla_e^2 + \frac{1}{m_p}\nabla_p^2 = \frac{1}{M}\nabla_R^2 + \frac{1}{m}\nabla_r^2$$

where $M = m_e + m_p$ is the total mass and $m = \frac{m_e m_p}{m_e + m_p}$ is the reduced mass.

Substitute this into the Schrödinger equation, which gives you

$$-\frac{\hbar^2}{2M}\nabla_R^2\psi(\boldsymbol{R}, \boldsymbol{r}) - \frac{\hbar^2}{2m}\nabla_r^2\psi(\boldsymbol{R}, \boldsymbol{r}) + V(\boldsymbol{r})\psi(\boldsymbol{R}, \boldsymbol{r}) = E\psi(\boldsymbol{R}, \boldsymbol{r})$$

And because $V(\boldsymbol{r}) = -\frac{e^2}{|\boldsymbol{r}_e - \boldsymbol{r}_p|}$, you get the following for the Schrödinger equation:

$$-\frac{\hbar^2}{2M}\nabla_R^2\psi(\mathbf{R}, \boldsymbol{r}) - \frac{\hbar^2}{2m}\nabla_r^2\psi(\mathbf{R}, \boldsymbol{r}) - \frac{e^2}{|\boldsymbol{r}_e - \boldsymbol{r}_p|}\psi(\mathbf{R}, \boldsymbol{r}) = E\psi(\mathbf{R}, \boldsymbol{r})$$

You know that $\boldsymbol{r} = \boldsymbol{r}_e - \boldsymbol{r}_p$, so write the potential as

$$V(\boldsymbol{r}) = -\frac{e^2}{|\boldsymbol{r}|}$$

Therefore, the Schrödinger equation becomes

$$-\frac{\hbar^2}{2M}\nabla_R^2\psi(\mathbf{R}, \boldsymbol{r}) - \frac{\hbar^2}{2m}\nabla_r^2\psi(\mathbf{R}, \boldsymbol{r}) - \frac{e^2}{|\boldsymbol{r}|}\psi(\mathbf{R}, \boldsymbol{r}) = E\psi(\mathbf{R}, \boldsymbol{r})$$

4 Separate the Schrödinger equation

$$-\frac{\hbar^2}{2M}\nabla_R^2\psi(\mathbf{R}, \boldsymbol{r}) - \frac{\hbar^2}{2m}\nabla_r^2\psi(\mathbf{R}, \boldsymbol{r}) - \frac{e^2}{|\boldsymbol{r}|}\psi(\mathbf{R}, \boldsymbol{r}) = E\psi(\mathbf{R}, \boldsymbol{r})$$

into two equations, one for $\boldsymbol{r}$ and one for $\mathbf{R}$, where $\mathbf{R} = \frac{m_e \boldsymbol{r}_e + m_p \boldsymbol{r}_p}{m_e + m_p}$ and $\boldsymbol{r} = \boldsymbol{r}_e - \boldsymbol{r}_p$.
The answers are

- $-\frac{\hbar^2}{2M}\nabla_R^2\psi(\mathbf{R}) = E_R\psi(\mathbf{R})$
- $-\frac{\hbar^2}{2m}\nabla_r^2\psi(\boldsymbol{r}) - \frac{e^2}{|\boldsymbol{r}|}\psi(\boldsymbol{r}) = E_r\psi(\boldsymbol{r})$, **where**
  - $\mathbf{R} = \frac{m_e r_e + m_p r_p}{m_e + m_p}$
  - $\boldsymbol{r} = \boldsymbol{r}_e - \boldsymbol{r}_p$
  - $M = m_e + m_p$
  - $m = \frac{m_e m_p}{m_e + m_p}$

The Schrödinger equation looks like this:

$$-\frac{\hbar^2}{2M}\nabla_R^2\psi(\mathbf{R},\boldsymbol{r})-\frac{\hbar^2}{2m}\nabla_r^2\psi(\mathbf{R},\boldsymbol{r})-\frac{e^2}{|\boldsymbol{r}|}\psi(\mathbf{R},\boldsymbol{r})=E\psi(\mathbf{R},\boldsymbol{r})$$

Because it contains terms involving either **R** or ***r*** but not both, it's a separable differential equation, which means there's a solution of the following form:

$$\psi(\mathbf{R},\boldsymbol{r}) = \psi(\mathbf{R})\psi(\boldsymbol{r})$$

Substitute $\psi(\mathbf{R},\boldsymbol{r}) = \psi(\mathbf{R})\psi(\boldsymbol{r})$ into the Schrödinger equation, giving you

$$-\frac{\hbar^2}{2M}\nabla_R^2\psi(\mathbf{R})\psi(\boldsymbol{r})-\frac{\hbar^2}{2m}\nabla_r^2\psi(\mathbf{R})\psi(\boldsymbol{r})-\frac{e^2}{|\boldsymbol{r}|}\psi(\mathbf{R})\psi(\boldsymbol{r})=E\psi(\mathbf{R})\psi(\boldsymbol{r})$$

Divide by $\psi(\mathbf{R})\psi(\boldsymbol{r})$ to get

$$-\frac{\hbar^2}{2M\psi(\mathbf{R})}\nabla_R^2\psi(\mathbf{R})-\frac{\hbar^2}{2m\psi(\boldsymbol{r})}\nabla_r^2\psi(\boldsymbol{r})-\frac{e^2}{|\boldsymbol{r}|}=E$$

The terms here depend on either $\psi(\mathbf{R})$ and $\psi(\boldsymbol{r})$ but not both. So you can separate Schrödinger equation into two equations, the first of which looks like this:

$$-\frac{\hbar^2}{2M\psi(\mathbf{R})}\nabla_R^2\psi(\mathbf{R})=E_{\mathbf{R}}$$

and the second of which looks like this:

$$-\frac{\hbar^2}{2m\psi(\boldsymbol{r})}\nabla_r^2\psi(\boldsymbol{r})-\frac{e^2}{|\boldsymbol{r}|}=E_r$$

where the total energy, E, equals $E_{\mathbf{R}} + E_r$. Multiply the first equation by $\psi(\mathbf{R})$, which gives you

$$-\frac{\hbar^2}{2M\psi}\nabla_R^2\psi(\mathbf{R})=E_{\mathbf{R}}\psi(\mathbf{R})$$

Then multiply the second equation by $\psi(\boldsymbol{r})$, which gives you

$$-\frac{\hbar^2}{2m}\nabla_r^2\psi(\boldsymbol{r})-\frac{e^2}{|\boldsymbol{r}|}\psi(\boldsymbol{r})=E_r\psi(\boldsymbol{r})$$

Now you have two Schrödinger equations, one for ***r*** and one for **R**.

5 The Schrödinger equation for $\psi(\boldsymbol{r})$ is

$$-\frac{\hbar^2}{2m}\nabla_r^2\psi(\boldsymbol{r})-\frac{e^2}{|\boldsymbol{r}|}\psi(\boldsymbol{r})=E_r\psi(\boldsymbol{r})$$

where $\boldsymbol{r} = \boldsymbol{r}_e - \boldsymbol{r}_p$. You can break the solution, $\psi(\boldsymbol{r})$, into a radial and an angular part:

$$\psi(\boldsymbol{r}) = R_{nl}(\boldsymbol{r})\, Y_{lm}(\theta, \phi)$$

The angular part of $\psi(\mathbf{r})$ is made up of spherical harmonics, $Y_{lm}(\theta, \phi)$. Now you have to solve for the radial part, $R_{nl}(\mathbf{r})$. The Schrödinger equation becomes this for the radial part:

$$-\frac{\hbar^2}{2m}\frac{d^2}{dr^2}r\,R_{nl}(r)+l(l+1)\frac{\hbar^2}{2mr^2}r\,R_{nl}(r)-\frac{e^2}{r}r\,R_{nl}(r)=E_r\,r\,R_{nl}(r)$$

where $r=|\mathbf{r}|$. Solve this equation for small r up to an arbitrary constant. **The answer is $R_{nl}(r) \approx Ar^l$ for small r.**

For small r, the radial wave function must vanish, so you have

$$-\frac{\hbar^2}{2m}\frac{d^2}{dr^2}\left(r\,R_{nl}(r)\right)+l(l+1)\frac{\hbar^2}{2mr^2}r\,R_{nl}(r)=0$$

Multiply by $2m/\hbar^2$:

$$-\frac{d^2}{dr^2}r\,R_{nl}(r)+\frac{l(l+1)}{r^2}r\,R_{nl}(r)=0$$

The solution to this differential equation looks like $R_{nl}(r) \approx Ar^l + Br^{-l-1}$. Note that $R_{nl}(r)$ must vanish as $r \to 0$ — but the r^{-l-1} term goes to infinity. That means that B must be 0, so you have this solution for small r:

$$R_{nl}(r) \approx Ar^l$$

6 The radial Schrödinger equation becomes the following for the radial part of $\psi(\mathbf{r})$:

$$-\frac{\hbar^2}{2m}\frac{d^2}{dr^2}r\,R_{nl}(r)+l(l+1)\frac{\hbar^2}{2mr^2}r\,R_{nl}(r)-\frac{e^2}{r}r\,R_{nl}(r)=E_r\,r\,R_{nl}(r)$$

where $\mathbf{r} = \mathbf{r}_e - \mathbf{r}_p$ and $r=|\mathbf{r}|$. Solve this equation for large r up to an arbitrary constant. **The answer is $R_{nl}(r) \approx Ae^{-\lambda r}$ for large r.**

For very large r, the Schrödinger equation becomes the following (to see this, just let r get very big in the Schrödinger equation given in the statement of this problem):

$$\frac{d^2}{dr^2}r\,R_{nl}(r)+\frac{2mE}{\hbar^2}r\,R_{nl}(r)=0$$

The electron is in a bound state in the hydrogen atom, so $E < 0$, and the solution is proportional to

$$R_{nl}(r) \sim Ae^{-\lambda r} + Be^{\lambda r}$$

where $\lambda = \frac{(-2mE)^{1/2}}{\hbar}$.

Note that this diverges as $r \to \infty$ because of the $Be^{\lambda r}$ term, so B must be equal to 0. That means that $R_{nl}(r) \approx Ae^{-\lambda r}$.

7 Given that the hydrogen atom wave functions look like this:

$$\psi_{nlm}(r,\theta,\phi)=\frac{\left(\frac{2}{nr_0}\right)^{3/2}\left((n-l-1)!\right)^{1/2}}{\left(2n(n+1)!\right)^{1/2}}e^{-r/(nr_0)}\left(\frac{2r}{nr_0}\right)^l L_{n-l-1}^{2l+1}\left(\frac{2r}{nr_0}\right)Y_{lm}(\theta,\phi)$$

where $r_0 = \frac{\hbar^2}{me^2}$ and where $L_{n-l-1}^{2l+1}(2r/(nr_0))$ is a generalized Laguerre polynomial:

- $L_0^b(r) = 1$
- $L_1^b(r) = -r + b + 1$
- $L_2^b(r) = -\frac{r^2}{6} - (b+2)r + \frac{(b+2)(b+1)}{2}$
- $L_3^b(r) = -\frac{r^3}{6} - \frac{(b+3)r^2}{2} - \frac{(b+2)(b+3)r}{2} + \frac{(b+1)(b+2)(b+3)}{6}$

find $\psi_{200}(r, \theta, \phi)$. **The answer is**

$$\boldsymbol{\psi_{200}(r, \theta, \phi) = \frac{1}{2^{1/2} r_0^{3/2}} e^{-r/(2r_0)} Y_{00}(\theta, \phi)}$$

where $\boldsymbol{r_0 = \frac{\hbar^2}{me^2}}$**.**

Here's what the wave function $\psi_{nlm}(r, \theta, \phi)$ looks like for hydrogen:

$$\psi_{nlm}(r, \theta, \phi) = \frac{\left(\frac{2}{nr_0}\right)^{3/2} ((n-l-1)!)^{1/2}}{(2n(n+1)!)^{1/2}} e^{-r/(nr_0)} \left(\frac{2r}{nr_0}\right)^l L_{n-l-1}^{2l+1}\left(\frac{2r}{nr_0}\right) Y_{lm}(\theta, \phi)$$

where $L_{n-l-1}^{2l+1}(2r/(nr_0))$ is a generalized Laguerre polynomial. Here are the first few generalized Laguerre polynomials:

- $L_0^b(r) = 1$
- $L_1^b(r) = -r + b + 1$
- $L_2^b(r) = -\frac{r^2}{6} - (b+2)r + \frac{(b+2)(b+1)}{2}$
- $L_3^b(r) = -\frac{r^3}{6} - \frac{(b+3)r^2}{2} - \frac{(b+2)(b+3)r}{2} + \frac{(b+1)(b+2)(b+3)}{6}$

Substitute this into the wave equation, which gives you $\psi_{200}(r, \theta, \phi)$:

$$\psi_{200}(r, \theta, \phi) = \frac{1}{2^{1/2} r_0^{3/2}} e^{-r/(2r_0)} Y_{00}(\theta, \phi)$$

8 Given that the hydrogen atom wave functions look like this:

$$\psi_{nlm}(r, \theta, \phi) = \frac{\left(\frac{2}{nr_0}\right)^{3/2} ((n-l-1)!)^{1/2}}{(2n(n+1)!)^{1/2}} e^{-r/(nr_0)} \left(\frac{2r}{nr_0}\right)^l L_{n-l-1}^{2l+1}\left(\frac{2r}{nr_0}\right) Y_{lm}(\theta, \phi)$$

where $r_0 = \frac{\hbar^2}{me^2}$ and where $L_{n-l-1}^{2l+1}\left(2r/(nr_0)\right)$ is a generalized Laguerre polynomial:

- $L_0^b(r) = 1$
- $L_1^b(r) = -r + b + 1$
- $L_2^b(r) = -\frac{r^2}{6} - (b+2)r + \frac{(b+2)(b+1)}{2}$
- $L_3^b(r) = -\frac{r^3}{6} - \frac{(b+3)r^2}{2} - \frac{(b+2)(b+3)r}{2} + \frac{(b+1)(b+2)(b+3)}{6}$

find $\psi_{300}(r, \theta, \phi)$. **The answer is**

$$\boldsymbol{\psi_{300}(r,\theta,\phi) = \frac{2}{3\left(3^{1/2}\right)r_0^{3/2}} e^{-r/(3r_0)} \left(1 - \frac{2r}{3r_0} + \frac{2r^2}{27r_0^2}\right) Y_{00}(\theta,\phi)}$$

where $\boldsymbol{r_0 = \frac{\hbar^2}{me^2}}$**.**

Here's what the wave function $\psi_{nlm}(r, \theta, \phi)$ looks like for hydrogen:

$$\psi_{nlm}(r,\theta,\phi) = \frac{\left(\frac{2}{nr_0}\right)^{3/2}\left((n-l-1)!\right)^{1/2}}{\left(2n(n+1)!\right)^{1/2}} e^{-r/(nr_0)} \left(\frac{2r}{nr_0}\right)^l L_{n-l-1}^{2l+1}\left(\frac{2r}{nr_0}\right) Y_{lm}(\theta,\phi)$$

where $L_{n-l-1}^{2l+1}\left(2r/(nr_0)\right)$ is a generalized Laguerre polynomial. Here are the first few Laguerre polynomials:

- $L_0^b(r) = 1$
- $L_1^b(r) = -r + b + 1$
- $L_2^b(r) = -\frac{r^2}{6} - (b+2)r + \frac{(b+2)(b+1)}{2}$
- $L_3^b(r) = -\frac{r^3}{6} - \frac{(b+3)r^2}{2} - \frac{(b+2)(b+3)r}{2} + \frac{(b+1)(b+2)(b+3)}{6}$

Substitute into the wave function to get $\psi_{300}(r, \theta, \phi)$:

$$\psi_{300}(r,\theta,\phi) = \frac{2}{3\left(3^{1/2}\right)r_0^{3/2}} e^{-r/(3r_0)} \left(1 - \frac{2r}{3r_0} + \frac{2r^2}{27r_0^2}\right) Y_{00}(\theta,\phi)$$

9 Solve for the energy levels of hydrogen in terms of the quantization number n.

The answer is $\boldsymbol{E_n = -\frac{me^4}{2n^2\hbar^2} = -\frac{e^2}{2r_0n^2} \quad n = 1, 2, 3, \ldots}$

where $\boldsymbol{r_0 = \frac{\hbar^2}{me^2}}$**.**

Start with the quantization condition found in the sample problem:

$$n\frac{(-2mE)^{1/2}}{\hbar} = \frac{me^2}{\hbar^2} \quad n = 1, 2, 3, \ldots$$

Square both sides of this equation to get

$$-n^2\frac{2mE}{\hbar^2}=\frac{m^2e^4}{\hbar^4} \quad n=1, 2, 3, \ldots$$

Divide both sides by m and multiply by $\hbar^2$:

$$-n^2 2E=\frac{me^4}{\hbar^2} \quad n=1, 2, 3, \ldots$$

Solve for the energy E, which gives you the following:

$$E=-\frac{me^4}{2n^2\hbar^2} \quad n=1, 2, 3, \ldots$$

Rename E as E_n because it depends on the principle quantum number n:

$$E_n=-\frac{me^4}{2n^2\hbar^2} \quad n=1, 2, 3, \ldots$$

This result is sometimes written in terms of the *Bohr radius* — the orbital radius that Niels Bohr calculated for the electron in a hydrogen atom, r_0. The Bohr radius is

$$r_0=\frac{\hbar^2}{me^2}$$

In terms of r_0, E_n equals

$$E_n=-\frac{me^4}{2n^2\hbar^2}=-\frac{e^2}{2r_0n^2} \quad n=1, 2, 3, \ldots$$

10 Find the first three energy levels of hydrogen numerically. **The answer is E = –13.6 eV when n = 1, E = –3.4 eV when n = 2, and E = –1.5 eV when n = 3.**

The energy levels of hydrogen are

$$E_n=-\frac{me^4}{2n^2\hbar^2} \quad n=1, 2, 3, \ldots$$

Find the ground state (where $n = 1$), the first excited state ($n = 2$), and the second excited state ($n = 3$). This energy is negative because the electron is in a bound state.

11 Verify that the angular momentum degeneracy of the $n = 1$ and $n = 2$ states is n^2. **For n = 1, degeneracy is 1; for n = 2, degeneracy = 4.**

For the ground state, $n = 1$, the degeneracy is one because l and therefore m can only equal zero for this state. For $n = 2$, you have these states therefore a degeneracy of four:

- $\psi_{200}(r, \theta, \phi)$
- $\psi_{21-1}(r, \theta, \phi)$
- $\psi_{210}(r, \theta, \phi)$
- $\psi_{211}(r, \theta, \phi)$

12 Find the degeneracy of the hydrogen atom when you consider spin in addition to angular momentum. **The answer is**

$$\textbf{Degeneracy} = \sum_{l=0}^{n-1} 2(2l+1) = 2n^2$$

The spin of the electron provides additional quantum states. When you add spin into the wave function, it becomes

$$\psi_{nl_m m_s}(r, \theta, \phi) = R_{nl}(r) Y_{lm}(\theta, \phi) |s, m_s\rangle$$

This wave function can take two different forms, depending on m_s, like this:

- $\psi_{nlm\frac{1}{2}}(r, s, \theta, \phi) = R_{nl}(r) Y_{lm}(\theta, \phi) |\frac{1}{2}, \frac{1}{2}\rangle$
- $\psi_{nlm-\frac{1}{2}}(r, s, \theta, \phi) = R_{nl}(r) Y_{lm}(\theta, \phi) |\frac{1}{2}, -\frac{1}{2}\rangle$

If you include the spin of the electron, there are two spin states for every state $|n\rangle$, so the degeneracy becomes the following:

$$\text{Degeneracy} = \sum_{l=0}^{n-1} 2(2l+1)$$

which equals $2n^2$.

Chapter 9

Corralling Many Particles Together

In This Chapter

- Understanding multiple-particle systems
- Working with identical and distinguishable particles
- Understanding fermions, bosons, and wave function symmetry

Creating wave functions for hydrogen atoms isn't necessarily easy (as I discuss in Chapter 8), so you can imagine how difficult it is to handle many particles at once — each of which can interact with the others.

This chapter focuses on systems of many identical particles and on systems of distinguishable but independent particles. Using identical particles makes solving problems easier because you don't have to keep track of which particle is which in terms of mass, spin, or other measurements. But even so, there's only so much you can do to find the wave function and energy levels of multiple identical particles, particularly if they interact with each other.

The problem is the interaction among the particles: When you have 15 charged particles, for example, you have to know the position of each one to know the resulting electric forces on every other particle. The distance between particles appears in the denominator of all the terms in the Hamiltonian because the electric potential is proportional to one over the distance. So solving such differential equations as a multi-particle Hamiltonian is nearly impossible. However, you can still say a surprising amount about multi-particle systems if you make a few approximations. This chapter helps you get a firmer grasp on how they work.

The 4-1-1 on Many-Particle Systems

A *many-particle system* is just that: a system of numerous particles. They could be in gaseous form, for example, and a wave function would have to keep track of the position of each particle. In this section, I start by taking a look at the wave function of a multi-particle system of identical particles.

In a system with identical particles, you keep track of each particle by its position, so the wave function looks like this:

$$\psi(\boldsymbol{r}_1, \boldsymbol{r}_2, \boldsymbol{r}_3, \ldots)$$

The following example and practice problems show you what else you can determine from general systems of identical multiple particles.

Q. What is the normalization condition for a general multiple-particle wave function $\psi(\mathbf{r}_1, \mathbf{r}_2, \mathbf{r}_3, ...)$, and what is the probability that particle 1 is in $d^3\mathbf{r}_1$, particle 2 is in $d^3\mathbf{r}_2$, particle 3 is in $d^3\mathbf{r}_3$, and so on?

A. $\int_{-\infty}^{+\infty} \left|\psi(\mathbf{r}_1, \mathbf{r}_2, \mathbf{r}_3 ...)\right|^2 d^3\mathbf{r}_1 d^3\mathbf{r}_2 d^3\mathbf{r}_3 ... = 1$ **and**

$\left|\psi(\mathbf{r}_1, \mathbf{r}_2, \mathbf{r}_3 ...)\right|^2 d^3\mathbf{r}_1 d^3\mathbf{r}_2 d^3\mathbf{r}_3 ...$

As for any wave function, the normalization of $\psi(\mathbf{r}_1, \mathbf{r}_2, \mathbf{r}_3, ...)$ demands that

$$\int_{-\infty}^{+\infty} \left|\psi(\mathbf{r}_1, \mathbf{r}_2, \mathbf{r}_3 ...)\right|^2 d^3\mathbf{r}_1 d^3\mathbf{r}_2 d^3\mathbf{r}_3 ... = 1$$

That's the normalization condition.

The probability that particle 1 is in $d^3\mathbf{r}_2$, particle 2 is in $d^3\mathbf{r}_2$, particle 3 is in $d^3\mathbf{r}_3$, and so on is given by the square of the wave function with the various positions equal to all the positions you want to find particle 1 at, particle 2 at, and so on; therefore, it looks like this:

$$\left|\psi(\mathbf{r}_1, \mathbf{r}_2, \mathbf{r}_3 ...)\right|^2 d^3\mathbf{r}_1 d^3\mathbf{r}_2 d^3\mathbf{r}_3 ...$$

1. What does the Hamiltonian look like for a general multiple-particle system?

Solve It

2. What does the total energy look like for a general multiple-particle system?

Solve It

Zap! Working with Multiple-Electron Systems

One type of multi-particle system is an atom with multiple electrons (there are other types of multi-electron systems, such as clouds of electrons, but atoms are most commonly used here). The electrons are identical and interchangeable. Consider the multiple-electron atom in Figure 9-1. Here, **R** is the coordinate of the nucleus, $\boldsymbol{r}_1$ is the coordinate of the first electron, $\boldsymbol{r}_2$ is the coordinate of the second electron, and so on. The figure shows three electrons, but in general, you give the total number of electrons as Z.

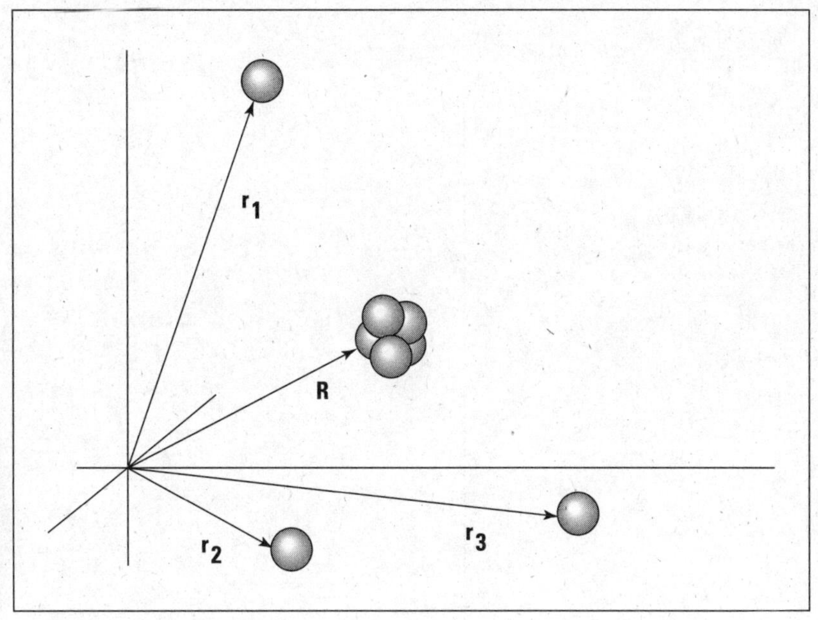

Figure 9-1: A multi-electron atom.

Although finding the wave function for such a system is difficult, you can still say some things about this system, such as how much total energy it has based on the number of electrons. The following problems show you how to do so.

Q. What is the total kinetic energy of the system in Figure 9-1, assuming that there are Z electrons?

A. $$\text{KE} = \sum_{i=1}^{Z} \frac{-\hbar^2 \nabla_i^2}{2m} - \frac{\hbar^2 \nabla_R^2}{2m}$$

The total kinetic energy of the electrons and the nucleus is just the sum of the individual kinetic energies.

3. What is the total potential energy of the system in Figure 9-1, assuming that there are Z electrons?

Solve It

4. What is the total energy of the system in Figure 9-1, assuming that there are Z electrons?

Solve It

The Old Shell Game: Exchanging Particles

One of the most powerful tools you have when working with multi-particle systems is to *exchange* the particles, or switch them with each other. Particular systems work one way when you exchange two particles, and other systems work in a different way; some allow exchange, and some don't, as this section shows. To see this in action, you can create an operator that exchanges two particles in a multi-particle system.

Consider the general wave function for N particles:

$$\psi(\boldsymbol{r}_1, \boldsymbol{r}_2, ..., \boldsymbol{r}_i, ..., \boldsymbol{r}_j, ..., \boldsymbol{r}_N)$$

Here's how to create an exchange operator, P_{ij}, that exchanges particles *i* and *j*. In other words

$$P_{ij}\psi(\boldsymbol{r}_1, \boldsymbol{r}_2, ..., \boldsymbol{r}_i, ..., \boldsymbol{r}_j, ..., \boldsymbol{r}_N) = \psi(\boldsymbol{r}_1, \boldsymbol{r}_2, ..., \boldsymbol{r}_j, ..., \boldsymbol{r}_i, ..., \boldsymbol{r}_N)$$

The following text takes a look at the exchange operator with an example and some practice problems.

Q. How does P_{ij} compare with P_{ji}?

A. $\mathbf{P}_{ij} = \mathbf{P}_{ji}$.

Start with the definition of the exchange operator P_{ij}:

$$P_{ij}\psi(\boldsymbol{r}_1, \boldsymbol{r}_2, ..., \boldsymbol{r}_i, ..., \boldsymbol{r}_j, ..., \boldsymbol{r}_N) = \psi(\boldsymbol{r}_1, \boldsymbol{r}_2, ..., \boldsymbol{r}_j, ..., \boldsymbol{r}_i, ..., \boldsymbol{r}_N)$$

Now write out the exchange operator for P_{ji}:

$$P_{ji}\psi(\boldsymbol{r}_1, r_2, ..., \boldsymbol{r}_i, ..., \boldsymbol{r}_j, ..., \boldsymbol{r}_N) = \psi(\boldsymbol{r}_1, \boldsymbol{r}_2, ..., \boldsymbol{r}_j, ..., \boldsymbol{r}_i, ..., \boldsymbol{r}_N)$$

Note that applying P_{ji} is the same as applying P_{ij}:

$$P_{ij}\psi(\boldsymbol{r}_1, \boldsymbol{r}_2, ..., \boldsymbol{r}_i, \boldsymbol{r}_j, ..., \boldsymbol{r}_i, ..., \boldsymbol{r}_N) = \psi(\boldsymbol{r}_1, \boldsymbol{r}_2, ..., ..., \boldsymbol{r}_j, ..., \boldsymbol{r}_N)$$
$$= P_{ji}\psi(\boldsymbol{r}_1, \boldsymbol{r}_2, ..., \boldsymbol{r}_i, ..., \boldsymbol{r}_j, ..., \boldsymbol{r}_N)$$

Therefore, $P_{ij} = P_{ji}$.

5. What is the result of applying P_{ij} twice?

Solve It

6. Show that if

$$\psi(r_1, r_2, r_3, r_4) = C\frac{r_1 e^{r_2}}{r_3 r_4}$$

where C is a constant, then P_{12} and P_{14} do not commute — that is, that $P_{12}P_{14}$ does not equal $P_{14}P_{12}$.

Solve It

Examining Symmetric and Antisymmetric Wave Functions

You can divide particles into those with symmetric and antisymmetric wave functions. Particles with symmetric wave functions are called *bosons,* and those with antisymmetric wave functions are called *fermions,* and they have quite different properties — especially in terms of sharing the same quantum numbers: Any number of bosons can have the same quantum numbers, but no two fermions may (that means no two fermions can occupy exactly the same state in a system).

This section takes a look at what it means to be symmetric or antisymmetric. Because $P_{ij}^2 = 1$, if a wave function is an eigenfunction of P_{ij}, then the possible eigenvalues are 1 and –1. In other words, for $\psi(\boldsymbol{r}_1, \boldsymbol{r}_2, ..., \boldsymbol{r}_i, ..., \boldsymbol{r}_j, ..., \boldsymbol{r}_N)$, an eigenfunction of P_{ij}, you can have one of the following situations:

- **Symmetric:** $P_{ij}\psi(\boldsymbol{r}_1, \boldsymbol{r}_2, ..., \boldsymbol{r}_i, ..., \boldsymbol{r}_j, ..., \boldsymbol{r}_N) = \psi(\boldsymbol{r}_1, \boldsymbol{r}_2, ..., \boldsymbol{r}_i, ..., \boldsymbol{r}_j, ..., \boldsymbol{r}_N)$
- **Antisymmetric:** $P_{ij}\psi(\boldsymbol{r}_1, \boldsymbol{r}_2, ..., \boldsymbol{r}_i, ..., \boldsymbol{r}_j, ..., \boldsymbol{r}_N) = -\psi(\boldsymbol{r}_1, \boldsymbol{r}_2, ..., \boldsymbol{r}_i, ..., \boldsymbol{r}_j, ..., \boldsymbol{r}_N)$

So there are two types of eigenfunctions of the exchange operator: symmetric and antisymmetric eigenfunctions. So how can you determine whether a wave function is symmetric or antisymmetric? The following example shows you how, and then I give you a few practice problems to experiment with.

Q. Is this wave function symmetric, antisymmetric, or neither?

$$\psi_3(\boldsymbol{r}_1, \boldsymbol{r}_2) = r_1 r_2 \frac{(r_1 - r_2)}{(r_1 - r_2)^2}$$

A. **Antisymmetric.** Here's how to check for symmetry:

1. **Apply the exchange operator to the wave function.**

   The wave function is

   $$\psi_3(\boldsymbol{r}_1, \boldsymbol{r}_2) = r_1 r_2 \frac{(r_1 - r_2)}{(r_1 - r_2)^2}$$

   Apply P_{12}:

   $$P_{12}\psi_3(\boldsymbol{r}_1, \boldsymbol{r}_2) = r_2 r_1 \frac{(r_2 - r_1)}{(r_2 - r_1)^2}$$

2. **Check how this answer compares to the original wave function.**

   You know that

   $$r_1 r_2 \frac{(r_2 - r_1)}{(r_2 - r_1)^2} = -r_2 r_1 \frac{(r_1 - r_2)}{(r_2 - r_1)^2}$$

   Therefore, $P_{12}\psi_3(\boldsymbol{r}_1, \boldsymbol{r}_2) = -\psi_3(\boldsymbol{r}_1, \boldsymbol{r}_2)$, so $\psi(\boldsymbol{r}_1, \boldsymbol{r}_2)$ is antisymmetric.

7. Is this wave function symmetric, anti-symmetric, or neither?

$$\psi(r_1, r_2) = \frac{r_1^{\,2} r_2^{\,2}}{(r_1 - r_2)^2}$$

Solve It

8. Is this wave function symmetric, anti-symmetric, or neither?

$$\psi(r_1, r_2) = (r_1 - r_2)^4$$

Solve It

9. Is this wave function symmetric, anti-symmetric, or neither?

$$\psi(r_1, r_2) = \frac{r_1^{\,2} + r_2^{\,2}}{r_2 r_1} + \frac{1}{(r_1 - r_2)^2}$$

Solve It

10. Is this wave function symmetric, anti-symmetric, or neither?

$$\psi(r_1, r_2) = \frac{r_1 r_2}{(r_1 - r_2)^2} + r_1^{\,2} - r_2^{\,2}$$

Solve It

Jumping into Systems of Many Distinguishable Particles

Some systems have many distinguishable particles. For example, each particle may have a different mass. Having a system of distinguishable particles means you can't just exchange two particles and get the same system back, so exchange operators are used less frequently here. However, you can still say something about such systems.

If the potential that each particle sees isn't dependent on the other particles (such as when you have charged distinguishable particles in an electric field), then this kind of system is solvable, and the math isn't too hard. Check out the following example and practice problems for more insight.

Q. What does the potential look like for a system of many independent distinguishable particles?

A. $PE = V(r_1, r_2, \ldots, r_N) = \sum_{i=1}^{N} V(r_i)$

If the particles are independent, the potential for all particles is just the sum of the individual potentials each particle sees, which looks like this summation, assuming there are N particles.

11. What is the Hamiltonian of a system of many independent distinguishable particles?

Solve It

12. What does the total wave function for a system of independent distinguishable particles look like in terms of the wave functions of each particle?

Solve It

Trapped in Square Wells: Many Distinguishable Particles

For a system of many distinguishable particles, you sometimes need to figure out what the bound states look like when the particles are trapped in a square well. Say you have four particles, each with a different mass, in a square well. And say the potential of the square well looks like this for each of the four noninteracting particles:

$$V_i(x_i) = \begin{cases} 0 & 0 \le x_i \le a \\ \infty & x_i > a \\ \infty & x_i < 0 \end{cases}$$

What do the wave functions and energy levels look like? The following example and practice problems help you make these determinations.

Q. What does the Schrödinger equation look like for four distinguishable independent particles in a square well when you divide it into four separate equations?

A. **Here are the equations:**

- $E_1\psi_{n_1}(x_1) = \frac{-\hbar^2}{2m_1}\frac{d^2}{dx_1^2}\psi_{n_1}(x_1)$
- $E_2\psi_{n_2}(x_2) = \frac{-\hbar^2}{2m_2}\frac{d^2}{dx_2^2}\psi_{n_2}(x_2)$
- $E_3\psi_{n_3}(x_3) = \frac{-\hbar^2}{2m_3}\frac{d^2}{dx_3^2}\psi_{n_3}(x_3)$
- $E_4\psi_{n_4}(x_4) = \frac{-\hbar^2}{2m_4}\frac{d^2}{dx_4^2}\psi_{n_4}(x_4)$

The Schrödinger equation looks like this:

$$E\psi_{n_1,n_2,n_3,n_4}(x_1, x_2, x_3, x_4) = \sum_{i=1}^{4}\frac{-\hbar^2}{2m_i}\frac{d^2}{dx_i^2}\psi_{n_1,n_2,n_3,n_4}(x_1, x_2, x_3, x_4)$$

Because each particle is independent, you can separate the Schrödinger equation into four one-particle equations, where i = 1, 2, 3, or 4.

13. What does the wave function look like for distinguishable particles in a square well?

Solve It

14. What are the energy levels of four distinguishable particles in a square well?

Solve It

Creating the Wave Functions of Symmetric and Antisymmetric Multi-Particle Systems

When you have two, three, or more identical particles, their wave function must be symmetric or antisymmetric — those are your only two choices; you can't have a wave function that's indeterminate with regard to symmetry. That means that only symmetric or antisymmetric wave functions are allowed. (For more on symmetry, see the earlier section titled "Examining Symmetric and Antisymmetric Wave Functions.")

Symmetric wave functions stay the same under particle exchange, and antisymmetric wave functions get a minus sign in front when you exchange two particles.

Take a look at how this works.

Q. What do the symmetric and antisymmetric wave functions of two free particles look like?

A. **Here are the answers:**

- $\psi_s(r_1s_1, r_2s_2) = \frac{1}{\sqrt{2}}\left[\psi_{n_1}(r_1s_1)\psi_{n_2}(r_2s_2) + \psi_{n_1}(r_2s_2)\psi_{n_2}(r_1s_1)\right]$
- $\psi_a(r_1s_1, r_2s_2) = \frac{1}{\sqrt{2}}\left[\psi_{n_1}(r_1s_1)\psi_{n_2}(r_2s_2) - \psi_{n_1}(r_2s_2)\psi_{n_2}(r_1s_1)\right]$

Here's the symmetric wave function, made up of single-particle wave functions:

$$\psi_s(r_1s_1, r_2s_2) = \frac{1}{\sqrt{2}}\left[\psi_{n_1}(r_1s_1)\psi_{n_2}(r_2s_2) + \psi_{n_1}(r_2s_2)\psi_{n_2}(r_1s_1)\right]$$

Here's the antisymmetric wave function, made up of the two single-particle wave functions; note that it changes sign under particle exchange:

$$\psi_a(r_1s_1, r_2s_2) = \frac{1}{\sqrt{2}}\left[\psi_{n_1}(r_1s_1)\psi_{n_2}(r_2s_2) - \psi_{n_1}(r_2s_2)\psi_{n_2}(r_1s_1)\right]$$

where n_i stands for all the quantum numbers of the ith particle.

Note that you can also write the symmetric wave function like this:

$$\psi_s(r_1s_1, r_2s_2) = \frac{1}{\sqrt{2}}\sum_{P} P\psi_{n_1}(r_1s_1)\psi_{n_2}(r_2s_2)$$

where P is the permutation operator, which takes the permutation of its argument. Similarly, you can write the antisymmetric wave function like this:

$$\psi_a(r_1s_1, r_2s_2) = \frac{1}{\sqrt{2}}\sum_{P}(-1)^{P} P\psi_{n_1}(r_1s_1)\psi_{n_2}(r_2s_2)$$

where the term $(-1)^P$ is 1 for even permutations (where you exchange both r_1s_1 and r_2s_2 and also n_1 and n_2) and –1 for odd permutations (where you exchange r_1s_1 and r_2s_2 but not n_1 and n_2, or n_1 and n_2 but not r_1s_1 and r_2s_2).

In fact, $\psi_a(r_1s_1, r_2s_2)$ is sometimes written in determinant form:

$$\psi_a(r_1s_1, r_2s_2) = \frac{1}{\sqrt{2!}}\begin{vmatrix}\psi_{n_1}(r_1s_1) & \psi_{n_1}(r_2s_2) \\ \psi_{n_2}(r_1s_1) & \psi_{n_2}(r_2s_2)\end{vmatrix}$$

15. What do the symmetric and antisymmetric wave functions of three free particles look like?

Solve It

16. What do the symmetric and antisymmetric wave functions of multiple free particles look like in general?

Solve It

Answers to Problems on Multiple-Particle Systems

Here are the answers to the practice questions I present earlier in this chapter, along with the original questions and answer explanations.

1 What does the Hamiltonian look like for a general multiple-particle system? **The answer is**

$$\mathbf{H}\psi(r_1, r_2, r_3, \ldots)$$
$$= \frac{p^2}{2m}\psi(r_1, r_2, r_3, \ldots) + \mathbf{V}(r)\psi(r_1, r_2, r_3, \ldots)$$
$$= \frac{-\hbar^2}{2m}\nabla^2\psi(r_1, r_2, r_3, \ldots) + \mathbf{V}(r)\psi(r_1, r_2, r_3, \ldots)$$

Start with the Hamiltonian:

$$H\psi(r_1, r_2, r_3, \ldots) = E\psi(r_1, r_2, r_3, \ldots)$$

When you're dealing only with a single particle (see Chapter 3), you can write this as

$$\frac{p^2}{2m}\psi(r) + V(r)\psi(r) = E\psi(r)$$

With multiple particles, you have to take into account all the particles for $H\psi(r_1, r_2, r_3, \ldots)$. This equals the following:

$$H\psi(r_1, r_2, r_3, \ldots)$$
$$= \frac{p^2}{2m}\psi(r_1, r_2, r_3, \ldots) + V(r)\psi(r_1, r_2, r_3, \ldots)$$
$$= \frac{-\hbar^2}{2m}\nabla^2\psi(r_1, r_2, r_3, \ldots) + V(r)\psi(r_1, r_2, r_3, \ldots)$$

2 What does the total energy look like for a general multiple-particle system? **The answer is**

$$\mathbf{H}\psi(r_1, r_2, r_3, \ldots) = \sum_{i=1}^{N}\frac{p_i^2}{2m_i}\psi(r_1, r_2, r_3, \ldots) + \mathbf{V}(r_1, r_2, r_3, \ldots)\psi(r_1, r_2, r_3, \ldots)$$
$$= \sum_{i=1}^{N}\frac{-\hbar^2}{2m_i}\nabla_i^2\psi(r_1, r_2, r_3, \ldots) + \mathbf{V}(r_1, r_2, r_3, \ldots)\psi(r_1, r_2, r_3, \ldots)$$

The total energy of a multiple-particle system is the sum of the energy of all the particles like this:

$$H\psi(r_1, r_2, r_3, \ldots) = \sum_{i=1}^{N}\frac{p_i^2}{2m_i}\psi(r_1, r_2, r_3, \ldots) + V(r_1, r_2, r_3, \ldots)\psi(r_1, r_2, r_3, \ldots)$$

And this equals

$$\sum_{i=1}^{N}\frac{-\hbar^2}{2m_i}\nabla_i^2\psi(r_1, r_2, r_3, \ldots) + V(r_1, r_2, r_3, \ldots)\psi(r_1, r_2, r_3, \ldots)$$

3 What is the total potential energy of the system in Figure 9-1, assuming that there are Z electrons?

$$\mathbf{PE} = \sum_{i=1}^{Z} \frac{Ze^2}{|\boldsymbol{r}_i - \mathbf{R}|} + \sum_{i>j} \frac{e^2}{|\boldsymbol{r}_i - \boldsymbol{r}_j|}$$

The potential energy of the multi-electron system is just the sum of the potential energies of the electrons (the second term) and the nucleus (the first term).

4 What is the total energy of the system in Figure 9-1, assuming that there are Z electrons?

E is the eigenvalue of

$$\sum_{i=1}^{Z} \frac{-\hbar^2}{2m_i}\nabla_i^2 - \frac{\hbar^2}{2m}\nabla_R^2 - \sum_{i=1}^{Z} \frac{Ze^2}{|\boldsymbol{r}_i - \mathbf{R}|} + \sum_{i>j} \frac{e^2}{|\boldsymbol{r}_i - \boldsymbol{r}_j|}$$

The total energy of a multiple-electron system is the sum of the total potential and kinetic energies.

5 What is the result of applying P_{ij} twice? **The answer is $P_{ij} P_{ij} = 1$.** Applying the exchange operator twice just puts the two exchanged particles back where they were originally, so $P_{ij}^2 = 1$. Here's what that looks like:

$$P_{ij} P_{ij} \psi(\boldsymbol{r}_1, \boldsymbol{r}_2, ..., \boldsymbol{r}_i, ..., \boldsymbol{r}_j, ..., \boldsymbol{r}_N) = P_{ij} \psi(\boldsymbol{r}_1, \boldsymbol{r}_2, ..., \boldsymbol{r}_j, ..., \boldsymbol{r}_i, ..., \boldsymbol{r}_N)$$
$$= \psi(\boldsymbol{r}_1, \boldsymbol{r}_2, ..., \boldsymbol{r}_i, ..., \boldsymbol{r}_j, ..., \boldsymbol{r}_N)$$

6 Show that if $\psi(\boldsymbol{r}_1, \boldsymbol{r}_2, \boldsymbol{r}_3, \boldsymbol{r}_4) = C\frac{r_1 e^{r_2}}{r_3 r_4}$ where C is a constant, then P_{12} and P_{14} do not commute — that is, that $P_{12}P_{14}$ does not equal $P_{14}P_{12}$. **The answer is**

$$\mathbf{P_{12} P_{14}} \psi(\boldsymbol{r}_1, \boldsymbol{r}_2, \boldsymbol{r}_3, \boldsymbol{r}_4) \neq \mathbf{P_{14} P_{12}} \psi(\boldsymbol{r}_1, \boldsymbol{r}_2, \boldsymbol{r}_3, \boldsymbol{r}_4)$$

The wave function is

$$\psi(\boldsymbol{r}_1, \boldsymbol{r}_2, \boldsymbol{r}_3, \boldsymbol{r}_4) = C\frac{r_1 e^{r_2}}{r_3 r_4}$$

Apply the exchange operator P_{14}, which looks like this:

$$P_{14}\psi(\boldsymbol{r}_1, \boldsymbol{r}_2, \boldsymbol{r}_3, \boldsymbol{r}_4) = C\frac{r_4 e^{r_2}}{r_3 r_1}$$

Then apply $P_{12} P_{14} \psi(r_1, r_2, r_3, r_4)$ like this:

$$P_{12}P_{14}\psi(\boldsymbol{r}_1, \boldsymbol{r}_2, \boldsymbol{r}_3, \boldsymbol{r}_4) = C\frac{r_4 e^{r_1}}{r_3 r_2}$$

Now look at $P_{14} P_{12} \psi(r_1, r_2, r_3, r_4)$, which looks like

$$P_{12}\psi(\boldsymbol{r}_1, \boldsymbol{r}_2, \boldsymbol{r}_3, \boldsymbol{r}_4) = C\frac{r_2 e^{r_1}}{r_3 r_4}$$

And here's what $P_{14} P_{12} \psi(r_1, r_2, r_3, r_4)$ is:

$$P_{14}P_{12}\psi\left(\boldsymbol{r}_1, \boldsymbol{r}_2, \boldsymbol{r}_3, \boldsymbol{r}_4\right) = C\frac{r_2 e^{r_4}}{r_3 r_1}$$

So $P_{12} P_{14} \psi(\boldsymbol{r}_1, \boldsymbol{r}_2, \boldsymbol{r}_3, \boldsymbol{r}_4) \neq P_{14} P_{12} \psi(\boldsymbol{r}_1, \boldsymbol{r}_2, \boldsymbol{r}_3, \boldsymbol{r}_4)$.

7 Is this wave function symmetric, antisymmetric, or neither?

$$\psi\left(\boldsymbol{r}_1, \boldsymbol{r}_2\right) = \frac{r_1^2 r_2^2}{\left(r_1 - r_2\right)^2}$$

Symmetric. The wave function is

$$\psi\left(\boldsymbol{r}_1, \boldsymbol{r}_2\right) = \frac{r_1^2 r_2^2}{\left(r_1 - r_2\right)^2}$$

Apply the exchange operator, P_{12}:

$$P_{12}\psi\left(\boldsymbol{r}_1, \boldsymbol{r}_2\right) = \frac{r_2^2 r_1^2}{\left(r_2 - r_1\right)^2}$$

See how these equations relate. Note that

$$\frac{r_2^2 r_1^2}{\left(r_2 - r_1\right)^2} = \frac{r_1^2 r_2^2}{\left(r_1 - r_2\right)^2}$$

$P_{12} \psi(\boldsymbol{r}_1, \boldsymbol{r}_2) = \psi(\boldsymbol{r}_1, \boldsymbol{r}_2)$, so $\psi(\boldsymbol{r}_1, \boldsymbol{r}_2)$ is symmetric.

8 Is this wave function symmetric, antisymmetric, or neither?

$$\psi(\boldsymbol{r}_1, \boldsymbol{r}_2) = (r_1 - r_2)^4$$

Symmetric. The wave function is

$$\psi(\boldsymbol{r}_1, \boldsymbol{r}_2) = (r_1 - r_2)^4$$

Apply the exchange operator P_{12}:

$$P_{12} \psi(\boldsymbol{r}_1, \boldsymbol{r}_2) = (r_2 - r_1)^4$$

Because $(r_1 - r_2)^4 = (r_2 - r_1)^4$, you know that $\psi(\boldsymbol{r}_1, \boldsymbol{r}_2)$ is a symmetric wave function, because $P_{12} \psi(\boldsymbol{r}_1, \boldsymbol{r}_2) = \psi(\boldsymbol{r}_1, \boldsymbol{r}_2)$.

9 Is this wave function symmetric, antisymmetric, or neither?

$$\psi\left(\boldsymbol{r}_1, \boldsymbol{r}_2\right) = \frac{r_1^2 + r_2^2}{r_2 r_1} + \frac{1}{\left(r_1 - r_2\right)^2}$$

Symmetric. The wave function is

$$\psi(\boldsymbol{r}_1, \boldsymbol{r}_2) = \frac{r_1^2 + r_2^2}{r_2 r_1} + \frac{1}{(r_1 - r_2)^2}$$

Apply the exchange operator P_{12}:

$$P_{12}\psi(\boldsymbol{r}_1, \boldsymbol{r}_2) = \frac{r_2^2 + r_1^2}{r_1 r_2} + \frac{1}{(r_2 - r_1)^2}$$

Note how the answer compares to the original wave function:

$$\frac{r_2^2 + r_1^2}{r_1 r_2} + \frac{1}{(r_2 - r_1)^2} = \frac{r_1^2 + r_2^2}{r_2 r_1} + \frac{1}{(r_1 - r_2)^2}$$

So $\psi(\boldsymbol{r}_1, \boldsymbol{r}_2)$ is symmetric.

10 Is this wave function symmetric, antisymmetric, or neither?

$$\psi(\boldsymbol{r}_1, \boldsymbol{r}_2) = \frac{r_1 r_2}{(r_1 - r_2)^2} + r_1^2 - r_2^2$$

Neither symmetric nor antisymmetric.

The wave function is

$$\psi(\boldsymbol{r}_1, \boldsymbol{r}_2) = \frac{r_1 r_2}{(r_1 - r_2)^2} + r_1^2 - r_2^2$$

Start by applying the exchange operator P_{12}:

$$P_{12}\psi(\boldsymbol{r}_1, \boldsymbol{r}_2) = \frac{r_2 r_1}{(r_2 - r_1)^2} + r_2^2 - r_1^2$$

How does $\psi(\boldsymbol{r}_1, \boldsymbol{r}_2)$ compare to $P_{12}\,\psi(\boldsymbol{r}_1, \boldsymbol{r}_2)$?

$$\frac{r_1 r_2}{(r_1 - r_2)^2} + r_1^2 - r_2^2 \neq \frac{r_2 r_1}{(r_2 - r_1)^2} + r_2^2 - r_1^2$$

So $\psi(\boldsymbol{r}_1, \boldsymbol{r}_2)$ is neither symmetric nor antisymmetric.

11 What is the Hamiltonian of a system of many independent distinguishable particles? **The answer is**

$$\mathbf{H}\psi(\boldsymbol{r}_1, \boldsymbol{r}_2, \ldots, \boldsymbol{r}_N) = \sum_{i=1}^{N}\left[\frac{-\hbar^2}{2m_i}\nabla_i^2 + \mathbf{V}_i(\boldsymbol{r}_i)\right]\psi(\boldsymbol{r}_1, \boldsymbol{r}_2, \ldots, \boldsymbol{r}_N)$$

You can cut the potential energy up into a sum of independent terms because the potential is independent of each particle, so here's what the Hamiltonian looks like:

$$H\psi(\boldsymbol{r}_1, \boldsymbol{r}_2, \ldots, \boldsymbol{r}_N) = \sum_{i=1}^{N}\left[\frac{-\hbar^2}{2m_i}\nabla_i^2 + V_i(\boldsymbol{r}_i)\right]\psi(\boldsymbol{r}_1, \boldsymbol{r}_2, \ldots, \boldsymbol{r}_N)$$

12 What does the total wave function for a system of independent distinguishable particles look like in terms of the wave functions of each particle? **The answer is**

$$\boldsymbol{\psi}_{n_1, n_2, \ldots, n_N}(\boldsymbol{r}_1, \boldsymbol{r}_2, \ldots, \boldsymbol{r}_N) = \prod_{i=1}^{N}\boldsymbol{\psi}_{n_i}(\boldsymbol{r}_i)$$

Because the particles are independent, the wave function is just the product of the individual wave functions, where the Π symbol is just like Σ, except that it stands for a product of terms.

13 What does the wave function look like for distinguishable particles in a square well? **The answer is**

$$\boldsymbol{\psi}_{n_1, n_2, n_3, n_4}(\boldsymbol{x}_1, \boldsymbol{x}_2, \boldsymbol{x}_3, \boldsymbol{x}_4) = \frac{\mathbf{4}}{\boldsymbol{a}^2}\,\mathbf{sin}\left(\frac{\boldsymbol{n}_1\boldsymbol{\pi}}{\boldsymbol{a}}\boldsymbol{x}_1\right)\mathbf{sin}\left(\frac{\boldsymbol{n}_2\boldsymbol{\pi}}{\boldsymbol{a}}\boldsymbol{x}_2\right)\mathbf{sin}\left(\frac{\boldsymbol{n}_3\boldsymbol{\pi}}{\boldsymbol{a}}\boldsymbol{x}_3\right)\mathbf{sin}\left(\frac{\boldsymbol{n}_4\boldsymbol{\pi}}{\boldsymbol{a}}\boldsymbol{x}_4\right)$$

For a one-dimensional system with a particle in a square well, the wave function is

$$\psi_i(x_i) = \frac{2^{\frac{1}{2}}}{a^{\frac{1}{2}}}\sin\left(\frac{n_i\pi}{a}x_i\right)$$

The wave function for a four-particle system is the product of the individual wave functions:

$$\psi_{n_1, n_2, n_3, n_4}(x_1, x_2, x_3, x_4) = \frac{4}{a^2}\sin\left(\frac{n_1\pi}{a}x_1\right)\sin\left(\frac{n_2\pi}{a}x_2\right)\sin\left(\frac{n_3\pi}{a}x_3\right)\sin\left(\frac{n_4\pi}{a}x_4\right)$$

As an example, in the ground state, $n_1 = n_2 = n_3 = n_4 = 1$, you have the following:

$$\psi_{1,1,1,1}(x_1, x_2, x_3, x_4) = \frac{4}{a^2}\sin\left(\frac{\pi}{a}x_1\right)\sin\left(\frac{\pi}{a}x_2\right)\sin\left(\frac{\pi}{a}x_3\right)\sin\left(\frac{\pi}{a}x_4\right)$$

14 What are the energy levels of four distinguishable particles in a square well? **The answer is**

$$\mathbf{E} = \frac{\boldsymbol{\hbar}^2\boldsymbol{\pi}^2}{\mathbf{2}\boldsymbol{a}^2}\left[\frac{\boldsymbol{n}_1^2}{\boldsymbol{m}_1} + \frac{\boldsymbol{n}_2^2}{\boldsymbol{m}_2} + \frac{\boldsymbol{n}_3^2}{\boldsymbol{m}_3} + \frac{\boldsymbol{n}_4^2}{\boldsymbol{m}_4}\right]$$

For one-particle square wells, the energy levels are

$$E_i = \frac{\hbar^2\pi^2 n_i^2}{2m_i a^2}$$

For a four-particle system, the total energy is the sum of the individual energies like this:

$$E = \sum_{i=1}^{4} E_i$$

So the energy is

$$E = \frac{\hbar^2\pi^2}{2a^2}\left[\frac{n_1^2}{m_1} + \frac{n_2^2}{m_2} + \frac{n_3^2}{m_3} + \frac{n_4^2}{m_4}\right]$$

For example, the energy of the ground state (that is, when all particles are in their ground state), where $n_1 = n_2 = n_3 = n_4 = 1$, is

$$E = \frac{\hbar^2\pi^2}{2a^2}\left[\frac{1}{m_1} + \frac{1}{m_2} + \frac{1}{m_3} + \frac{1}{m_4}\right]$$

15 What do the symmetric and antisymmetric wave functions of three free particles look like? **The symmetric wave function looks like this:**

$$\psi_s\left(r_1s_1,\, r_2s_2,\, r_3s_3\right) = \frac{1}{\sqrt{3!}}\sum_P P\psi_{n_1}\left(r_1s_1\right)\psi_{n_2}\left(r_2s_2\right)\psi_{n_3}\left(r_3s_3\right)$$

And the antisymmetric wave function looks like this:

$$\psi_a\left(r_1s_1,\, r_2s_2,\, r_2s_2\right) = \frac{1}{\sqrt{3!}}\sum_P (-1)^P P\psi_{n_1}\left(r_1s_1\right)\psi_{n_2}\left(r_2s_2\right)\psi_{n_3}\left(r_3s_3\right)$$

When you have three free particles, the symmetric wave function just adds the three wave functions:

$$\psi_s\left(r_1s_1,\, r_2s_2,\, r_3s_3\right) = \frac{1}{\sqrt{3!}}\sum_P P\psi_{n_1}\left(r_1s_1\right)\psi_{n_2}\left(r_2s_2\right)\psi_{n_3}\left(r_3s_3\right)$$

The antisymmetric wave function includes alternating minus signs:

$$\psi_a\left(r_1s_1,\, r_2s_2,\, r_3s_3\right) = \frac{1}{\sqrt{3!}}\sum_P (-1)^P P\psi_{n_1}\left(r_1s_1\right)\psi_{n_2}\left(r_2s_2\right)\psi_{n_3}\left(r_3s_3\right)$$

16 What do the symmetric and antisymmetric wave functions of multiple free particles look like in general? **The symmetric wave function looks like this:**

$$\psi_s\left(r_1s_1,\, r_2s_2,\, \ldots,\, r_Ns_N\right) = \frac{1}{\sqrt{N!}}\sum_P P\psi_{n_1}\left(r_1s_1\right)\psi_{n_2}\left(r_2s_2\right),\, \ldots,\, \psi_{n_N}\left(r_Ns_N\right)$$

And the antisymmetric wave function looks like this:

$$\psi_a\left(r_1s_1,\, r_2s_2,\, \ldots,\, r_Ns_N\right) = \frac{1}{\sqrt{N!}}\sum_P (-1)^P P\psi_{n_1}\left(r_1s_1\right)\psi_{n_2}\left(r_2s_2\right),\, \ldots,\, \psi_{n_N}\left(r_Ns_N\right)$$

For a system of N particles, the symmetric wave function looks like this — just the sum of the wave functions (where N! is N factorial):

$$\psi_s\left(r_1s_1,\, r_2s_2,\, \ldots,\, r_Ns_N\right) = \frac{1}{\sqrt{N!}}\sum_P P\psi_{n_1}\left(r_1s_1\right)\psi_{n_2}\left(r_2s_2\right),\, \ldots,\, \psi_{n_N}\left(r_Ns_N\right)$$

The antisymmetric wave function looks like this, with alternating minus signs:

$$\psi_a\left(r_1s_1,\, r_2s_2,\, \ldots,\, r_Ns_N\right) = \frac{1}{\sqrt{N!}}\sum_P (-1)^P P\psi_{n_1}\left(r_1s_1\right)\psi_{n_2}\left(r_2s_2\right),\, \ldots,\, \psi_{n_N}\left(r_Ns_N\right)$$

Part IV

Acting on Impulse — Impacts in Quantum Physics

In this part . . .

This part introduces you to working with perturbation theory and scattering theory. Perturbation theory gives systems a push and then predicts what's going to happen so you can describe more-complex situations. Scattering theory is all about what happens when one particle hits another — at what angle will the particles separate? With what momentums? It's all coming up in this part.

Chapter 10

Pushing with Perturbation Theory

In This Chapter

- Using perturbation theory to make slight adjustments
- Applying electric fields to harmonic oscillators

Quantum physics allows you to solve only basic systems, such as free particles, square wells, harmonic oscillators, and so on. Yet the real world has all kinds of systems, and of course many don't match the ideal-world systems that are readily solvable.

That's where perturbation theory comes into the picture. *Perturbation theory* lets you mix two different types of systems — as long as the addition you're making to a known system is small. For example, you may have a charged particle oscillating in a harmonic oscillator kind of way, and then you add a constant electric field to the system. If that constant electric field is weak compared to the harmonic oscillator potential, then you can use perturbation theory to find the new energy levels and the new wave functions. This chapter takes a closer look at how this theory works and provides some practice problems for you to build your skills.

Examining Perturbation Theory with Energy Levels and Wave Functions

Perturbation theory proceeds by a series of approximations, which is why the perturbation to a known system must be small; otherwise, the small-order approximations become too large to be accurate. Basically, perturbation theory allows you to approximate the solution more and more accurately. This section points out how to use perturbation theory to solve problems with energy levels and wave functions.

When working with this theory, you start with a Hamiltonian for the unperturbed system. H_0 is a known Hamiltonian, with known eigenfunctions and eigenvalues. You then add a small addition due to the *perturbing effect* (for example, an electric field), λW, where $\lambda \ll 1$. The λW is the so-called *perturbation Hamiltonian,* where $\lambda \ll 1$ indicates that the perturbation Hamiltonian is small. Here's the equation:

$$H = H_0 + \lambda W \quad (\lambda \ll 1)$$

Determining the eigenstates of the Hamiltonian in $H = H_0 + \lambda W$ (where $\lambda \ll 1$) is what solving problems like this is all about. In other words, here's the problem you want to solve, the Hamiltonian applied to a wave vector:

$$H|\psi_n\rangle = (H_0 + \lambda W)|\psi_n\rangle = E_n|\psi_n\rangle \quad (\lambda \ll 1)$$

The following example and practice problems examine this theory with energy levels and the wave functions. Later in this section, you solve the perturbed Schrödinger equation for the first- and second-order corrections.

Q. Express a perturbed system's energy levels in terms of an expansion in λ.

A. $E_n = E_n^{(0)} + \lambda E_n^{(1)} + \lambda^2 E_n^{(2)} + \ldots \quad (\lambda \ll 1)$

Here's how to solve the problem:

1. **Start with the energy of the unperturbed system.**

   $E_n = E_n^{(0)} + \ldots$

2. **Add the first-order correction to the energy, $\lambda E_n^{(1)}$.**

   $E_n = E_n^{(0)} + \lambda E_n^{(1)} + \ldots \qquad (\lambda \ll 1)$

3. **Add the second-order correction to the energy as well, $\lambda^2 E_n^{(2)}$.**

   $E_n = E_n^{(0)} + \lambda E_n^{(1)} + \lambda^2 E_n^{(2)} + \ldots \qquad (\lambda \ll 1)$

1. Proceeding by analogy with the example problem, what does an expansion of the wave function of the perturbed system look like, expressed as an expansion in λ (to the second order)?

Solve It

2. What does the perturbed Hamiltonian look like when you multiply it by the perturbed wave function to give you the perturbed Schrödinger equation?

Solve It

Solving the perturbed Schrödinger equation for the first-order correction

In this section, I start with solving for the first-order correction. The *first-order correction* is the first approximation to the solution of the perturbed Schrödinger equation, and finding it is the first step to solving the Schrödinger equation.

Here's the perturbed Schrödinger equation (see problem 2 for the calculations):

$$\left(H_0+\lambda W\right)\left(\left|\phi_n\right\rangle+\lambda\left|\psi_n^{(1)}\right\rangle+\lambda^2\left|\psi_n^{(2)}\right\rangle+\ldots\right)$$
$$=\left(E_n^{(0)}+\lambda E_n^{(1)}+\lambda^2 E_n^{(2)}+\ldots\right)\left(\left|\phi_n\right\rangle+\lambda\left|\psi_n^{(1)}\right\rangle+\lambda^2\left|\psi_n^{(2)}\right\rangle+\ldots\right)\quad(\lambda\ll 1)$$

You solve the perturbed Schrödinger equation by noting that the coefficients of λ must all be equal:

- The zeroth-order term in λ gives you the following equation:

  $$H_0\left|\phi_n\right\rangle=E_n^{(0)}\left|\phi_n\right\rangle$$

  You use this zeroth-order equation to solve perturbation problems.

- For the first-order terms in λ, equating them gives you

  $$H_0\left|\psi_n^{(1)}\right\rangle+W\left|\phi_n\right\rangle=E_n^{(0)}\left|\psi_n^{(1)}\right\rangle+E_n^{(1)}\left|\phi_n\right\rangle$$

  That's the first-order equation you use to solve perturbation problems.

- Next you equate the coefficients of λ^2, the second-order terms, giving you

  $$H_0\left|\psi_n^{(2)}\right\rangle+W\left|\psi_n^{(1)}\right\rangle=E_n^{(0)}\left|\psi_n^{(2)}\right\rangle+E_n^{(1)}\left|\psi_n^{(1)}\right\rangle+E_n^{(2)}\left|\phi_n\right\rangle$$

  That's the equation you derive from the second order in λ.

Now you have to solve for $E_n^{(1)}$ and $E_n^{(2)}$ using the zeroth-, first-, and second-order perturbation equations. The following example shows you how, and the practice problems allow you to find the solutions on your own.

Q. Given that $\left\langle\phi_n\middle|\psi_n\right\rangle=1$, where the unperturbed wave function is $\left|\phi_n\right\rangle$ and the perturbed wave function is $\left|\psi_n\right\rangle$, show that $\left\langle\phi_n\middle|\psi_n^{(1)}\right\rangle=\left\langle\phi_n\middle|\psi_n^{(2)}\right\rangle=0$, where $\left|\psi_n^{(1)}\right\rangle$ and $\left|\psi_n^{(2)}\right\rangle$ are the first- and second-order corrections to the perturbed wave function.

A. $\boldsymbol{\left\langle\phi_n\middle|\psi_n^{(1)}\right\rangle=0}$ **and** $\boldsymbol{\left\langle\phi_n\middle|\psi_n^{(2)}\right\rangle=0}$**.**

The unperturbed wave function, $\left|\phi_n\right\rangle$, isn't very different from the perturbed wave function, $\left|\psi_n\right\rangle$, because the perturbation is small. That means that

$$\left\langle\phi_n\middle|\psi_n\right\rangle\approx 1$$

In fact, you can normalize $|\psi_n\rangle$ so that $\langle\phi_n|\psi_n\rangle$ is exactly equal to 1, as assumed in the statement of this equation:

$$\langle\phi_n|\psi_n\rangle = 1$$

Given that

$$|\psi_n\rangle = |\phi_n\rangle + \lambda\left|\psi_n^{(1)}\right\rangle + \lambda^2\left|\psi_n^{(2)}\right\rangle + \ldots \quad (\lambda \ll 1)$$

you get the following:

$$\lambda\left\langle\phi_n\middle|\psi_n^{(1)}\right\rangle + \lambda^2\left\langle\phi_n\middle|\psi_n^{(2)}\right\rangle + \ldots = 0$$

Because the coefficients of λ must both vanish (that is, λ is not zero), you get the following:

- $\left\langle\phi_n\middle|\psi_n^{(1)}\right\rangle = 0$
- $\left\langle\phi_n\middle|\psi_n^{(2)}\right\rangle = 0$

---

3. Proceeding by analogy with the example problem, what does an expansion of the wave function look like? Solve for the first-order correction to the energy for the perturbed system, $E_n^{(1)}$, in terms of W, the perturbation Hamiltonian, starting with the first-order equation you use to solve perturbation problems:

$$H_0\left|\psi_n^{(1)}\right\rangle + W|\phi_n\rangle = E_n^{(0)}\left|\psi_n^{(1)}\right\rangle + E_n^{(1)}|\phi_n\rangle$$

Solve It

4. Multiply $\left|\psi_n^{(1)}\right\rangle$ by the following expression, which equals 1:

$$\sum_m |\phi_m\rangle\langle\phi_m|$$

to solve for $\left|\psi_n^{(1)}\right\rangle$, the first-order correction to wave function.

Solve It

Solving the perturbed Schrödinger equation for the second-order correction

Now I turn to the *second-order correction* to the solution to the Schrödinger equation, which is the next approximation after the first-order correction. Finding the second-order correction and adding it to the first gives you an even more accurate solution to the Schrödinger equation.

This section takes a look at finding the second-order correction to the energy levels. Here I explain how you can find $E_n^{(2)}$. For small perturbations, finding the first- and second-order corrections to the energy levels should be mathematically accurate enough for most purposes.

Q. Find the second-order correction to the energy levels, $E_n^{(2)}$, in terms of $\left|\psi_n^{(1)}\right\rangle$.

A. $\mathbf{E_n^{(2)} = \langle \phi_n | W | \psi_n^{(1)} \rangle}$

Start by multiplying both sides of

$$H_0\left|\psi_n^{(2)}\right\rangle + W\left|\psi_n^{(1)}\right\rangle = E_n^{(0)}\left|\psi_n^{(2)}\right\rangle + E_n^{(1)}\left|\psi_n^{(1)}\right\rangle + E_n^{(2)}|\phi_n\rangle$$

by $\langle\phi_n|$ to get the following:

$$\left\langle\phi_n\middle|H_0\middle|\psi_n^{(2)}\right\rangle + \left\langle\phi_n\middle|W\middle|\psi_n^{(1)}\right\rangle = \left\langle\phi_n\middle|E_n^{(0)}\middle|\psi_n^{(2)}\right\rangle + \left\langle\phi_n\middle|E_n^{(1)}\middle|\psi_n^{(1)}\right\rangle + \left\langle\phi_n\middle|E_n^{(2)}\middle|\phi_n\right\rangle$$

Note that $\left\langle\phi_n\middle|\psi_n^{(1)}\right\rangle$ is equal to zero, so the second term on the right drops out. You get

$$\left\langle\phi_n\middle|H_0\middle|\psi_n^{(2)}\right\rangle + \left\langle\phi_n\middle|W\middle|\psi_n^{(1)}\right\rangle = \left\langle\phi_n\middle|E_n^{(0)}\middle|\psi_n^{(2)}\right\rangle + \left\langle\phi_n\middle|E_n^{(2)}\middle|\phi_n\right\rangle$$

Because $\left\langle\phi_n\middle|\psi_n^{(2)}\right\rangle$ is also equal to zero, you get

$$\left\langle\phi_n\middle|W\middle|\psi_n^{(1)}\right\rangle = \left\langle\phi_n\middle|E_n^{(2)}\middle|\phi_n\right\rangle$$

Because $E_n^{(2)}$ is just a number, you have

$$\left\langle\phi_n\middle|W\middle|\psi_n^{(1)}\right\rangle = E_n^{(2)}\langle\phi_n|\phi_n\rangle$$

And because $\langle\phi_n|\phi_n\rangle = 1$, you have the following:

$$E_n^{(2)} = \left\langle\phi_n\middle|W\middle|\psi_n^{(1)}\right\rangle$$

5. Convert $E_n^{(2)} = \left\langle \phi_n \middle| W \middle| \psi_n^{(1)} \right\rangle$ to a form involving the unperturbed wave functions, $|\phi_n\rangle$, instead of the first-order correction to the wave function, $\left|\psi_n^{(1)}\right\rangle$.

Solve It

6. Find the total energy of a perturbed system according to perturbation theory, including the first- and second-order corrections.

Solve It

Applying Perturbation Theory to the Real World

Are you ready to use some numbers? How about seeing the effect of perturbing a harmonic oscillator? This section gives you an example of perturbation theory so you can see how it applies to real life.

Say that you have a small particle oscillating in a harmonic potential, back and forth. The Hamiltonian looks like this:

$$H = -\frac{\hbar^2}{2m}\frac{d^2}{dx^2} + \frac{1}{2}m\omega^2 x^2$$

where the particle's mass is m, its location is x, and the angular frequency of the motion is ω. (See Chapter 3 for an introduction to harmonic oscillators.)

Next, assume that the particle is charged, with charge q, and that you apply a weak electric field, ε, to the particle. The force due to the electric field in this case is the perturbation, and the Hamiltonian becomes the following:

$$H = -\frac{\hbar^2}{2m}\frac{d^2}{dx^2} + \frac{1}{2}m\omega^2 x^2 + q\varepsilon x$$

What are the energies of this Hamiltonian? Those are the allowed energy levels of the oscillating system. I take a look at solving this equation by using perturbation theory.

Q. What are the energy levels of the perturbed Hamiltonian?

$$H = -\frac{\hbar^2}{2m}\frac{d^2}{dx^2} + \frac{1}{2}m\omega^2 x^2 + q\varepsilon x$$

Solve for the energy levels exactly.

A. $\mathbf{E}_n = \left(n + \frac{1}{2}\right)\hbar\omega - \frac{q^2\varepsilon^2}{2m\omega^2}$

You can solve for the exact energy eigenvalues by making the substitution

$$y = x + \frac{q\varepsilon}{m\omega^2}$$

Solve for x:

$$x = y - \frac{q\varepsilon}{m\omega^2}$$

Substitute this new form of x into the Hamiltonian, which gives you the following:

$$H = -\frac{\hbar^2}{2m}\frac{d^2}{dy^2} + \frac{1}{2}m\omega^2 y^2 - \frac{q^2\varepsilon^2}{2m\omega^2}$$

Note that the last term is a constant:

$$H = -\frac{\hbar^2}{2m}\frac{d^2}{dy^2} + \frac{1}{2}m\omega^2 y^2 + C$$

where $C = -\frac{q^2\varepsilon^2}{2m\omega^2}$. This is the Hamiltonian of a harmonic oscillator with an added constant, which means that the energy levels are

$$E_n = \left(n + \frac{1}{2}\right)\hbar\omega + C$$

Substitute in for C, which gives you the exact energy levels:

$$E_n = \left(n + \frac{1}{2}\right)\hbar\omega - \frac{q^2\varepsilon^2}{2m\omega^2}$$

7. Use perturbation theory to find the energy levels of the harmonic oscillator with an applied electric field where the Hamiltonian is

$$H = -\frac{\hbar^2}{2m}\frac{d^2}{dx^2} + \frac{1}{2}m\omega^2 x^2 + q\varepsilon x$$

Solve It

8. Find the wave functions for the particle in harmonic oscillation with an added electric field, where this is the Hamiltonian:

$$H = -\frac{\hbar^2}{2m}\frac{d^2}{dx^2} + \frac{1}{2}m\omega^2 x^2 + q\varepsilon x$$

Include the first-order correction.

Solve It

Answers to Problems on Perturbation Theory

Here are the answers and explanations to the practice questions I present earlier in this chapter.

1 Proceeding by analogy with the example problem, what does an expansion of the wave function of the perturbed system look like, expressed as an expansion in λ (to the second order)? **The answer is** $\boldsymbol{|\psi_n\rangle = |\phi_n\rangle + \lambda|\psi_n^{(1)}\rangle + \lambda^2|\psi_n^{(2)}\rangle + \ldots \quad (\lambda \ll 1)}$**.**

Start with the wave function of the unperturbed system, $|\phi_n\rangle$:

$$|\psi_n\rangle = |\phi_n\rangle + \ldots$$

Add the first-order correction, $\lambda|\psi_n^{(1)}\rangle$:

$$|\psi_n\rangle = |\phi_n\rangle + \lambda|\psi_n^{(1)}\rangle + \ldots \quad (\lambda \ll 1)$$

Then add to that the second-order correction to the wave function, $\lambda^2|\psi_n^{(2)}\rangle$:

$$|\psi_n\rangle = |\phi_n\rangle + \lambda|\psi_n^{(1)}\rangle + \lambda^2|\psi_n^{(2)}\rangle + \ldots \quad (\lambda \ll 1)$$

2 What does the perturbed Hamiltonian look like when you multiply it by the perturbed wave function to give you the perturbed Schrödinger equation? **The answer is**

$$\begin{aligned}&\mathbf{(H_0 + \lambda W)\left(|\phi_n\rangle + \lambda|\psi_n^{(1)}\rangle + \lambda^2|\psi_n^{(2)}\rangle + \ldots\right)}\\ &\mathbf{= \left(E_n^{(0)} + \lambda E_n^{(1)} + \lambda^2 E_n^{(2)} + \ldots\right)\left(|\phi_n\rangle + \lambda|\psi_n^{(1)}\rangle + \lambda^2|\psi_n^{(2)}\rangle + \ldots\right) \quad (\lambda \ll 1)}\end{aligned}$$

Start with the perturbed Hamiltonian:

$$H|\psi_n\rangle = (H_0 + \lambda W)|\psi_n\rangle = E_n|\psi_n\rangle \quad (\lambda \ll 1)$$

Plug in the perturbed wave function:

$$|\psi_n\rangle = |\phi_n\rangle + \lambda|\psi_n^{(1)}\rangle + \lambda^2|\psi_n^{(2)}\rangle + \ldots \quad (\lambda \ll 1)$$

Recall that the perturbed energy looks like this:

$$E_n = E_n^{(0)} + \lambda E_n^{(1)} + \lambda^2 E_n^{(2)} + \ldots \quad (\lambda \ll 1)$$

Putting all those equations together gives you the following:

$$\begin{aligned}&(H_0 + \lambda W)\left(|\phi_n\rangle + \lambda|\psi_n^{(1)}\rangle + \lambda^2|\psi_n^{(2)}\rangle + \ldots\right)\\ &= \left(E_n^{(0)} + \lambda E_n^{(1)} + \lambda^2 E_n^{(2)} + \ldots\right)\left(|\phi_n\rangle + \lambda|\psi_n^{(1)}\rangle + \lambda^2|\psi_n^{(2)}\rangle + \ldots\right) \quad (\lambda \ll 1)\end{aligned}$$

3 Proceeding by analogy with the example problem, what does an expansion of the wave function look like? Solve for the first-order correction to the energy for the perturbed system, $E_n^{(1)}$, in terms of W, the perturbation Hamiltonian, starting with the first-order equation you use to solve perturbation problems:

$$H_0\left|\psi_n^{(1)}\right\rangle + W\left|\phi_n\right\rangle = E_n^{(0)}\left|\psi_n^{(1)}\right\rangle + E_n^{(1)}\left|\phi_n\right\rangle$$

The answer is $E_n^{(1)} = \langle\phi_n|W|\phi_n\rangle$.

Start with the first-order equation you use to solve perturbation problems:

$$H_0\left|\psi_n^{(1)}\right\rangle + W\left|\phi_n\right\rangle = E_n^{(0)}\left|\psi_n^{(1)}\right\rangle + E_n^{(1)}\left|\phi_n\right\rangle$$

Multiply by $\langle\phi_n|$ to get the following:

$$\left\langle\phi_n\left|H_0\right|\psi_n^{(1)}\right\rangle + \left\langle\phi_n\left|W\right|\phi_n\right\rangle = \left\langle\phi_n\left|E_n^{(0)}\right|\psi_n^{(1)}\right\rangle + \left\langle\phi_n\left|E_n^{(1)}\right|\phi_n\right\rangle$$

Simplify this by subtracting the first term on each side to get the equal terms

$$E_n^{(1)} = \left\langle\phi_n\left|W\right|\phi_n\right\rangle$$

4 Multiply $\left|\psi_n^{(1)}\right\rangle$ by the following expression, which equals 1:

$$\sum_m \left|\phi_m\right\rangle\left\langle\phi_m\right|$$

to solve for $\left|\psi_n^{(1)}\right\rangle$, the first-order correction to wave function. **The answer is**

$$\left|\psi_n^{(1)}\right\rangle = \sum_{m\neq n} \frac{\langle\phi_m|W|\phi_n\rangle}{E_n^{(0)} - E_m^{(0)}}\left|\phi_m\right\rangle$$

Multiply $\left|\psi_n^{(1)}\right\rangle$ by the following expression, which is equal to 1:

$$\sum_m \left|\phi_m\right\rangle\left\langle\phi_m\right|$$

Here's what you get:

$$\left|\psi_n^{(1)}\right\rangle = \left(\sum_m \left|\phi_m\right\rangle\left\langle\phi_m\right|\right)\left|\psi_n^{(1)}\right\rangle$$

And that equation equals the following:

$$\left|\psi_n^{(1)}\right\rangle = \sum_{m\neq n} \left\langle\phi_m\middle|\psi_n^{(1)}\right\rangle\left|\phi_m\right\rangle$$

The $m = n$ term is zero because $\left\langle\phi_n\middle|\psi_n^{(1)}\right\rangle = 0$.

You can find $\left\langle \phi_m \middle| \psi_n^{(1)} \right\rangle$ by multiplying $\langle \phi_m |$ by the first-order correction, which follows:

$$H_0 \left| \psi_n^{(1)} \right\rangle + W|\phi_n\rangle = E_n^{(0)} \left| \psi_n^{(1)} \right\rangle + E_n^{(1)} |\phi_n\rangle$$

Here's what you get:

$$\left\langle \phi_m \middle| \psi_n^{(1)} \right\rangle = \frac{\langle \phi_m | W | \phi_n \rangle}{E_n^{(0)} - E_m^{(0)}}$$

Substitute this into your equation for $\left| \psi_n^{(1)} \right\rangle$, which gives you

$$\left| \psi_n^{(1)} \right\rangle = \sum_{m \neq n} \frac{\langle \phi_m | W | \phi_n \rangle}{E_n^{(0)} - E_m^{(0)}} |\phi_m\rangle$$

That's the first-order correction to the wave function, $\left| \psi_n^{(1)} \right\rangle$. Therefore, the wave function of the perturbed system, to the first order, is

$$|\psi_n\rangle = |\phi_n\rangle + \sum_{m \neq n} \frac{\langle \phi_m | \lambda W | \phi_n \rangle}{E_n^{(0)} - E_m^{(0)}} |\phi_m\rangle \ldots \quad (\lambda << 1)$$

5 Convert $E_n^{(2)} = \left\langle \phi_n \middle| W \middle| \psi_n^{(1)} \right\rangle$ to a form involving the unperturbed wave functions, $|\phi_n\rangle$, instead of the first-order correction to the wave function, $\left| \psi_n^{(1)} \right\rangle$. **The answer is**

$$\mathbf{E_n^{(2)} = \sum_{m \neq n} \frac{\left| \langle \phi_m | W | \phi_n \rangle \right|^2}{E_n^{(0)} - E_m^{(0)}}}$$

The problem asks you to start with

$$E_n^{(2)} = \left\langle \phi_n \middle| W \middle| \psi_n^{(1)} \right\rangle$$

From problem 4, you know that

$$\left| \psi_n^{(1)} \right\rangle = \sum_{m \neq n} \frac{\langle \phi_m | W | \phi_n \rangle}{E_n^{(0)} - E_m^{(0)}} |\phi_m\rangle$$

Substitute that value of $\left| \phi_n^{(1)} \right\rangle$ into the equation for $E_n^{(2)}$, which gives you

$$E_n^{(2)} = \left\langle \phi_n \middle| W \middle| \psi_n^{(1)} \right\rangle$$

$$= \langle \phi_n | W \sum_{m \neq n} \frac{\langle \phi_m | W | \phi_n \rangle}{E_n^{(0)} - E_m^{(0)}} |\phi_m\rangle$$

Combining terms gives you

$$E_n^{(2)} = \sum_{m \neq n} \frac{\left| \langle \phi_m | W | \phi_n \rangle \right|^2}{E_n^{(0)} - E_m^{(0)}}$$

6 Find the total energy of a perturbed system according to perturbation theory, including the first- and second-order corrections. **The answer is**

$$\mathbf{E}_n = \mathbf{E}_n^{(0)} + \lambda\langle\phi_n|\mathbf{W}|\phi_n\rangle + \lambda^2\sum_{m\neq n}\frac{|\langle\phi_m|\mathbf{W}|\phi_n\rangle|^2}{\mathbf{E}_n^{(0)}-\mathbf{E}_m^{(0)}} + \ldots \quad (\lambda \ll 1)$$

You've already solved for $E_n^{(1)}$ and $E_n^{(2)}$ in problems 3 and 5:

$$E_n^{(1)} = \langle\phi_n|W|\phi_n\rangle$$

$$E_n^{(2)} = \sum_{m\neq n}\frac{|\langle\phi_m|W|\phi_n\rangle|^2}{E_n^{(0)}-E_m^{(0)}}$$

The total energy with the first- and second-order corrections is

$$E_n = E_n^{(0)} + \lambda E_n^{(1)} + \lambda^2 E_n^{(2)} + \ldots \qquad (\lambda \ll 1)$$

Substitute in the values of $E_n^{(1)}$ and $E_n^{(2)}$. The total energy is

$$E_n = E_n^{(0)} + \lambda\langle\phi_n|W|\phi_n\rangle + \lambda^2\sum_{m\neq n}\frac{|\langle\phi_m|W|\phi_n\rangle|^2}{E_n^{(0)}-E_m^{(0)}} + \ldots \quad (\lambda \ll 1)$$

7 Use perturbation theory to find the energy levels of the harmonic oscillator with an applied electric field where the Hamiltonian is

$$H = -\frac{\hbar^2}{2m}\frac{d^2}{dx^2} + \frac{1}{2}m\omega^2x^2 + q\varepsilon x$$

The answer is

$$\mathbf{E}_n = \left(n+\frac{1}{2}\right)\hbar\omega - \frac{q^2\varepsilon^2}{2m\omega^2}$$

The corrected energy is given by

$$E_n = E_n^{(0)} + \lambda\langle\phi_n|W|\phi_n\rangle + \lambda^2\sum_{m\neq n}\frac{|\langle\phi_m|W|\phi_n\rangle|^2}{E_n^{(0)}-E_m^{(0)}} + \ldots \quad (\lambda \ll 1)$$

where λW is the perturbation term in the Hamiltonian. That is, here, λW equals $q\varepsilon x$. The first-order correction is

$$\begin{aligned}\lambda\langle\phi_n|W|\phi_n\rangle &= \langle\phi_n|q\varepsilon x|\phi_n\rangle\\ &= q\varepsilon\langle\phi_n|x|\phi_n\rangle\end{aligned}$$

Note that $\langle\phi_n|x|\phi_n\rangle = 0$, because that's the expectation value of x, and for harmonic oscillators, the average value of x is zero. So the first-order correction to the energy, as given by perturbation theory, is zero.

So what is the second-order correction to the energy? It's

$$\lambda^2 \sum_{m \neq n} \frac{\left|\langle \phi_m | W | \phi_n \rangle\right|^2}{E_n^{(0)} - E_m^{(0)}} \quad (\lambda \ll 1)$$

Because $\lambda W = q\varepsilon x$, you get the following:

$$q^2\varepsilon^2 \sum_{m \neq n} \frac{\left|\langle \phi_m | x | \phi_n \rangle\right|^2}{E_n^{(0)} - E_m^{(0)}}$$

Changing notation from $\langle \phi_m |$ to $\langle m |$ and $| \phi_n \rangle$ to $| n \rangle$ to gives you the following:

$$q^2\varepsilon^2 \sum_{m \neq n} \frac{\left|\langle m | x | n \rangle\right|^2}{E_n^{(0)} - E_m^{(0)}}$$

Take this expression apart term by term. The zeroth order energy is

$$E_n^{(0)} = \left(n + \frac{1}{2}\right)\hbar\omega$$

For a harmonic oscillator, the following equations are true:

- $\langle n+1 | x | n \rangle = (n+1)^{1/2} \dfrac{(\hbar)^{1/2}}{(2m\omega)^{1/2}}$
- $\langle n-1 | x | n \rangle = (n)^{1/2} \dfrac{(\hbar)^{1/2}}{(2m\omega)^{1/2}}$

In addition, $E_n^{(0)} - E_{n-1}^{(0)} = \hbar\omega$ and $E_n^{(0)} - E_{n+1}^{(0)} = -\hbar\omega$. So make the appropriate substitutions. The second-order correction becomes

- $q^2\varepsilon^2 \dfrac{\left|\langle n+1 | x | n \rangle\right|^2}{E_n^{(0)} - E_{n+1}^{(0)}} + \ldots$
- $q^2\varepsilon^2 \dfrac{\left|\langle n-1 | x | n \rangle\right|^2}{E_n^{(0)} - E_{n-1}^{(0)}} + \ldots$

Substitute in the for $E_n^{(0)} - E_{n+1}^{(0)}$ and $E_n^{(0)} - E_{n-1}^{(0)}$, which gives you the following:

- $q^2\varepsilon^2 \dfrac{\left|\langle n+1 | x | n \rangle\right|^2}{-\hbar\omega} + \ldots$
- $q^2\varepsilon^2 \dfrac{\left|\langle n-1 | x | n \rangle\right|^2}{\hbar\omega} + \ldots$

Now substitute for $\langle n+1|x|n\rangle$ and $\langle n-1|x|n\rangle$, which gives you

- $q^2\varepsilon^2 \dfrac{(n+1)\hbar}{(-\hbar\omega)(2m\omega)} + \ldots$
- $q^2\varepsilon^2 \dfrac{n\hbar}{(\hbar\omega)(2m\omega)} + \ldots$

Simplify. The $\hbar$ cancels out, so

- $-q^2\varepsilon^2 \dfrac{(n+1)}{2m\omega^2} + \ldots$
- $q^2\varepsilon^2 \dfrac{n}{2m\omega^2} + \ldots$

Therefore, adding these terms together, the second-order correction is

$$-\frac{q^2\varepsilon^2}{2m\omega^2}\ldots$$

Thus, the energy of the harmonic oscillator in the electric field should be

$$E_n = \left(n+\frac{1}{2}\right)\hbar\omega - \frac{q^2\varepsilon^2}{2m\omega^2}$$

Note that perturbation theory gives you the same result as the exact answer in this case.

8 Find the wave functions for the particle in harmonic oscillation with an added electric field, where this is the Hamiltonian:

$$H = -\frac{\hbar^2}{2m}\frac{d^2}{dx^2} + \frac{1}{2}m\omega^2 x^2 + q\varepsilon x$$

Include the first-order correction. **Here's the answer:**

$$|\psi_n\rangle = |n\rangle + \frac{q\varepsilon}{\hbar\omega}\frac{\hbar^{1/2}}{(2m\omega)^{1/2}}\left(n^{1/2}|n-1\rangle - (n+1)^{1/2}|n+1\rangle\right)$$

The corrected wave function including the first-order correction is

$$|\psi_n\rangle = |\phi_n\rangle + \sum_{m\neq n}\frac{\langle\phi_m|\lambda W|\phi_n\rangle}{E_n^{(0)} - E_m^{(0)}}|\phi_m\rangle \ldots \quad (\lambda \ll 1)$$

Change the notation to use $\langle n|$ and $|n\rangle$; this becomes

$$|\psi_n\rangle = |n\rangle + \sum_{m\neq n}\frac{\langle m|\lambda W|n\rangle}{E_n^{(0)} - E_m^{(0)}}|m\rangle \ldots \quad (\lambda \ll 1)$$

And because $\lambda W = q\varepsilon x$, you can substitute for λW. The equation becomes

$$|\psi_n\rangle = |n\rangle + q\varepsilon\sum_{m\neq n}\frac{\langle m|x|n\rangle}{E_n^{(0)} - E_m^{(0)}}|m\rangle \ldots \quad (\lambda \ll 1)$$

In fact, only two terms are nonzero because $\langle n|x|n\rangle = 0$. The two nonzero terms are

- $\langle n+1|x|n\rangle = (n+1)^{1/2}\dfrac{(\hbar)^{1/2}}{(2m\omega)^{1/2}}$
- $\langle n-1|x|n\rangle = (n)^{1/2}\dfrac{(\hbar)^{1/2}}{(2m\omega)^{1/2}}$

And because the following equations are true:

- $E_n^{(0)} - E_{n-1}^{(0)} = \hbar\omega$
- $E_n^{(0)} - E_{n+1}^{(0)} = -\hbar\omega$

you know that

$$|\psi_n\rangle = |n\rangle + \frac{q\varepsilon}{\hbar\omega}\frac{\hbar^{1/2}}{(2m\omega)^{1/2}}\left(n^{1/2}|n-1\rangle - (n+1)^{1/2}|n+1\rangle\right)$$

In other words, adding an electric field to a harmonic oscillator spreads the wave function of the harmonic oscillator so that it includes adjacent states.

Chapter 11

One Hits the Other: Scattering Theory

In This Chapter

- Finding differential cross sections
- Changing between lab and center-of-mass frames
- Finding scattering amplitude and putting the Born approximation to work

In quantum physics, *scattering theory* has to do with — you guessed it — microscopic particles hitting each other. Two particles come at each other at great speed and — wham! — hit each other and then depart, usually at great speed. Quantum physics has a lot to say about scattering theory, and you solve problems using its approach in this chapter.

Cross Sections: Experimenting with Scattering

To completely understand scattering theory, you need to know how a scattering experiment works. To get an idea of what a scattering experiment looks like, look at Figure 11-1.

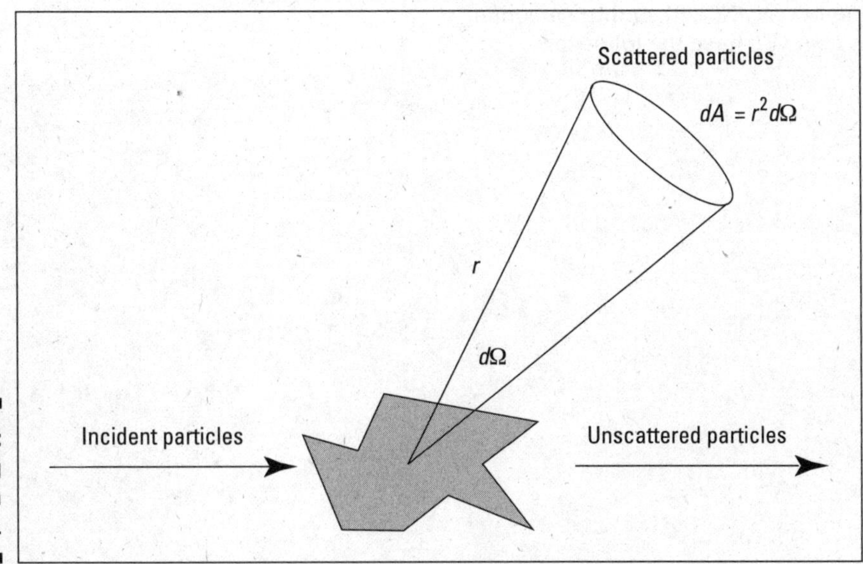

Figure 11-1: Scattering from a target.

As you can see, particles are being sent in a stream from the left and are interacting with a target. Most of them continue on unscattered, but some particles interact with the target and scatter. The particles that do scatter do so at a particular angle. You give the scattering angle as a solid angle, $d\Omega$, which equals $\sin\theta\, d\theta\, d\phi$, where ϕ and θ are the spherical angles (which I introduce in Chapter 7).

The *total cross section,* σ, is the cross section for scattering of any kind, through any angle.

The number of particles scattered into a specific $d\Omega$ per unit time is proportional to a very important quantity in scattering theory: the differential cross section. In quantum physics, the *differential cross section* is given by $d\sigma(\phi, \theta)/d\Omega$. It's a measure of the number of particles per second scattered into $d\Omega$ per incoming flux. The *incident flux,* J, (also called the *current density*) is the number of incident particles per unit area per unit time.

I take a look at the differential cross section and total cross section in the following example and practice problems.

Q. Relate $d\sigma(\phi, \theta)/d\Omega$ to $dN(\phi, \theta)/d\Omega$.

A. $$\frac{d\sigma(\phi,\theta)}{d\Omega} = \frac{1}{J}\frac{dN(\phi,\theta)}{d\Omega}$$

Start with the differential cross section:

$$\frac{d\sigma(\phi,\theta)}{d\Omega}$$

The differential cross section is the number of particles per second scattered into $d\Omega$ per incoming flux. You can write the number as $dN(\phi, \theta)$, and the incoming flux is J, so you have the following:

$$\frac{d\sigma(\phi,\theta)}{d\Omega} = \frac{1}{J}\frac{dN(\phi,\theta)}{d\Omega}$$

where $N(\phi, \theta)$ is the number of particles at angles ϕ and θ.

Tip: Note that $d\sigma(\phi, \theta)/d\Omega$ has the dimensions of area, so calling it a *cross section* is appropriate. Think of it as the size of the bull's-eye when you're aiming to scatter incident particles through a specific solid angle.

1. How do you relate the total cross section, σ, to the differential cross section, $d\sigma(\phi, \theta)/d\Omega$, in terms of Ω?

Solve It

2. How do you relate the total cross section, σ, to the differential cross section, $d\sigma(\phi, \theta)/d\Omega$, in terms of θ and ϕ?

Solve It

A Frame of Mind: Going from the Lab Frame to the Center-of-Mass Frame

Scattering experiments occur in the lab, but you do scattering calculations in the center-of-mass frame. Using the center-of-mass frame is a lot easier because the net momentum is zero in that frame.

Figure 11-2 shows a scattering in the *lab frame.* The first particle, traveling at v_{1lab}, is incident on another particle ($v_{2lab} = 0$) and hits it. After the collision, the first particle is scattered at angle θ_1, traveling at v'_{1lab}, and the other particle is scattered at angle θ_2 and velocity v'_{2lab}.

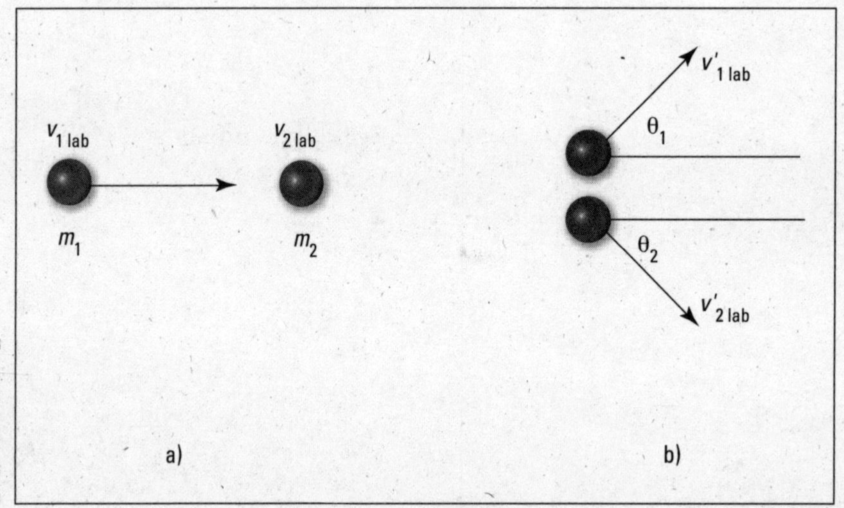

Figure 11-2: Scattering in the lab frame.

Take a look at the same scattering in the *center-of-mass frame* (where the center of mass of the entire system is stationary) in Figure 11-3. In the center-of-mass frame, the particles are heading toward each other. After they collide, they head away from each other at angles θ and $\pi - \theta$. Giving each particle the same momentum (in opposite directions) makes the calculations easier.

Much scattering theory calculations involve translating between the lab and center-of-mass frames, so this section looks at how to translate between those frames in a nonrelativistic way. For example, how do you relate the angles θ_1 (from the center-of-mass frame) and θ (from the lab frame)? Take a look at the example and the next few problems.

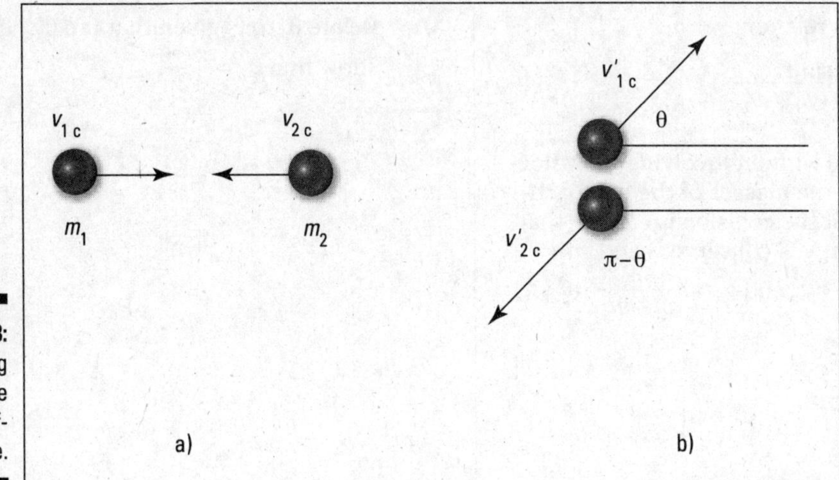

Figure 11-3: Scattering in the center-of-mass frame.

EXAMPLE

Q. How do you relate the angles θ_1 (from the center-of-mass frame) and θ (from the lab frame) in terms of v_{cm} and v'_{1c}?

A. $\tan\theta_1 = \dfrac{\sin\theta}{\cos\theta + v_{cm}/v'_{1c}}$

Note that you can connect v_{1lab} and v_{1c} using the velocity of the center of mass, v_{cm}, this way:

$$\mathbf{v}_{1lab} = \mathbf{v}_{1c} + \mathbf{v}_{cm}$$

And you can relate the velocity of particle 1 after its collision with particle 2 this way:

$$\mathbf{v}'_{1lab} = \mathbf{v}'_{1c} + \mathbf{v}_{cm}$$

Find the components of these velocities by using cos θ and sin θ:

- $v'_{1lab}\cos\theta_1 = v'_{1c}\cos\theta + v_{cm}$
- $v'_{1lab}\sin\theta_1 = v'_{1c}\sin\theta$

Divide these two equations by each other, as such:

$$\tan\theta_1 = \frac{\sin\theta}{\cos\theta + v_{cm}/v'_{1c}}$$

3. Start with this relation:

$$\tan\theta_1 = \frac{\sin\theta}{\cos\theta + v_{cm}/v'_{1c}}$$

Relate θ_1 and θ without involving the velocities; use only the masses of the two particles, given that the collision is *elastic* (that is, kinetic energy is conserved) and $\boldsymbol{v}_{cm} = \frac{m_1}{m_1+m_2}\boldsymbol{v}_{1lab}$ and $\boldsymbol{v}_{1c} = \frac{m_2}{m_1+m_2}\boldsymbol{v}_{1lab}$.

Solve It

4. Relate θ_2 to θ; given that $\tan\theta_2 = \cot(\theta/2)$, show that $\theta_2 = \frac{\pi-\theta}{2}$.

Solve It

Target Practice: Taking Cross Sections from the Lab Frame to the Center-of-Mass Frame

The differential cross section is $d\sigma/d\Omega$ (see the earlier section "Cross sections: Experimenting with Scattering" for details). The differential $d\sigma$ is infinitesimal in size, and it stays the same between the lab frame and the center-of-mass frame. But what about $d\Omega$? The angles that make up $d\Omega$ differ when you translate between frames. This section takes a look at how those angles differ.

Here's how the differential cross sections relate:

- **Lab frame:** $d\Omega_1 = \sin\theta_1\, d\theta_1\, d\phi_1$ for the lab differential cross section:

$$\left.\frac{d\sigma(\phi,\theta)}{d\Omega}\right|_{lab}$$

- **Center-of-mass frame:** $d\Omega = \sin\theta\, d\theta\, d\phi$ for the center-of-mass differential cross section:

$$\left.\frac{d\sigma(\phi,\theta)}{d\Omega}\right|_{cm}$$

The following example and practice problems look at how that works.

Q. Show that the following is true:

$$\left.\frac{d\sigma(\phi,\theta)}{d\Omega_1}\right|_{\text{lab}} = \left.\frac{d\sigma(\phi,\theta)}{d\Omega}\right|_{\text{cm}} \frac{d(\cos\theta)}{d(\cos\theta_1)}$$

A. $$\left.\frac{d\sigma(\phi,\theta)}{d\Omega_1}\right|_{\text{lab}} = \left.\frac{d\sigma(\phi,\theta)}{d\Omega}\right|_{\text{cm}} \frac{d(\cos\theta)}{d(\cos\theta_1)}$$

In the lab, $d\Omega_1 = \sin\theta_1\, d\theta_1\, d\phi_1$. And in the center-of-mass frame, $d\Omega = \sin\theta\, d\theta\, d\phi$. Because $d\sigma_{\text{lab}} = d\sigma_{\text{cm}}$, you get

$$\left.\frac{d\sigma(\phi,\theta)}{d\Omega_1}\right|_{\text{lab}} d\Omega_1 = \left.\frac{d\sigma(\phi,\theta)}{d\Omega}\right|_{\text{cm}} d\Omega$$

Put these three equations together, which gives you

$$\left.\frac{d\sigma(\phi,\theta)}{d\Omega_1}\right|_{\text{lab}} = \left.\frac{d\sigma(\phi,\theta)}{d\Omega}\right|_{\text{cm}} \frac{\sin\theta}{\sin\theta_1}\frac{d\theta}{d\theta_1}\frac{d\phi}{d\phi_1}$$

Because the angles are symmetric, $\phi = \phi_1$, so

$$\left.\frac{d\sigma(\phi,\theta)}{d\Omega_1}\right|_{\text{lab}} = \left.\frac{d\sigma(\phi,\theta)}{d\Omega}\right|_{\text{cm}} \frac{\sin\theta}{\sin\theta_1}\frac{d\theta}{d\theta_1}$$

And you can write this as

$$\left.\frac{d\sigma(\phi,\theta)}{d\Omega_1}\right|_{\text{lab}} = \left.\frac{d\sigma(\phi,\theta)}{d\Omega}\right|_{\text{cm}} \frac{d(\cos\theta)}{d(\cos\theta_1)}$$

---

5. Start with the following relation:

$$\left.\frac{d\sigma(\phi,\theta)}{d\Omega_1}\right|_{\text{lab}} = \left.\frac{d\sigma(\phi,\theta)}{d\Omega}\right|_{\text{cm}} \frac{d(\cos\theta)}{d(\cos\theta_1)}$$

Convert it to this, adding the masses:

$$\left.\frac{d\sigma(\phi,\theta)}{d\Omega_1}\right|_{\text{lab}} = \frac{\left(1+\frac{m_1^2}{m_2^2}+2\cos\theta\left(\frac{m_1}{m_2}\right)\right)^{3/2}}{1+\cos\theta\left(\frac{m_1}{m_2}\right)} \left.\frac{d\sigma(\phi,\theta)}{d\Omega}\right|_{\text{cm}}$$

Solve It

6. Start with two particles of equal mass colliding in the lab frame, where one particle starts at rest. Show that the two particles end up traveling at right angles with respect to each other in the lab frame.

Solve It

Getting the Goods on Elastic Scattering

Sometimes during a scattering experiment, the scattering of two particles is *elastic* — kinetic energy is conserved. When kinetic energy is conserved, you can say much more about the results of the scattering.

In this instance of an elastic scattering, assume that the interaction between the particles depends only on their relative distance, $|\mathbf{r}_1 - \mathbf{r}_2|$. In this section, you start by getting the Schrödinger equation. Then you can use that equation to solve for the probability that a particle is scattered into a solid angle $d\Omega$ — this probability is given by the differential cross section, $d\sigma/d\Omega$.

Q. Find the Schrödinger equation for the incident and scattered particle system.

A. $\frac{-\hbar^2}{2\mu}\nabla^2\psi(r)+\mathbf{V}(r)\psi(r)=\mathbf{E}\psi(r)$, where $\mu=\frac{m_1 m_2}{m_1+m_2}$.

You can change problems of this kind into two decoupled problems. The first decoupled equation considers the center of mass of the two particles as a free particle, and the second equation is for a fictitious particles of mass $m_1m_2/(m_1 + m_2)$. The first equation isn't significant when you're discussing scattering. The second equation is the one to concentrate on.

7. Find the wave function of an incident particle. Assume that the scattering potential $V(\mathbf{r})$ has a very finite range and find the wave function outside that range.

Solve It

8. The wave function of a scattered particle looks like this:

$$\phi_{sc}(\mathbf{r}) = Af(\phi, \theta)\frac{e^{i\mathbf{k}\cdot\mathbf{r}}}{r}$$

where $f(\phi, \theta)$ is the scattering amplitude, A is a dimensionless normalization factor, and $\left|\mathbf{k}^2\right| = \frac{2\mu E}{\hbar^2}$. Find the dimensions of $f(\phi, \theta)$.

Solve It

The Born Approximation: Getting the Scattering Amplitude of Particles

The scattering amplitude is $f(\phi, \theta)$, where the scattered wave looks like this:

$$\phi_{sc}(r) = Af(\phi, \theta)\frac{e^{ik\cdot r}}{r}$$

A is a normalization factor, and $\left|k^2\right| = 2\mu E/\hbar^2$.

Relating the scattering amplitude to the differential cross section is easy (for info on the differential cross section, see the earlier section titled "Cross Sections: Experimenting with Scattering"):

$$\frac{d\sigma(\phi, \theta)}{d\Omega} = \left|f(\phi, \theta)\right|^2$$

So as soon as you find the scattering amplitude, you've found the differential cross section. To find the scattering amplitude, you have to solve the Schrödinger equation:

$$\frac{-\hbar^2}{2\mu}\nabla^2\psi(\mathbf{r}) + V(\mathbf{r})\psi(\mathbf{r}) = E\psi(\mathbf{r})$$

You can also write this as

$$\left(\nabla^2 + k^2\right)\psi(\boldsymbol{r}) = \frac{2\mu}{\hbar^2}V(\boldsymbol{r})\psi(\boldsymbol{r})$$

You can express the solution to this differential equation as the sum of a homogeneous solution and a particular solution:

$$\psi(\boldsymbol{r}) = \psi_h(\boldsymbol{r}) + \psi_p(\boldsymbol{r})$$

The homogeneous solution satisfies the following equation:

$$\left(\nabla^2 + k^2\right)\psi_h(\boldsymbol{r}) = 0$$

So the homogeneous solution is a plane wave, corresponding to the incident plane wave:

$$\psi_h(\boldsymbol{r}) = Ae^{i\boldsymbol{k}_0\cdot\boldsymbol{r}}$$

You can find the particular solution in terms of *Green's functions,* so the solution to the Schrödinger equation is

$$\psi(\boldsymbol{r}) = Ae^{\boldsymbol{k}_0\cdot\boldsymbol{r}} + \frac{2\mu}{\hbar^2}\int G(\boldsymbol{r}-\boldsymbol{r}')V(\boldsymbol{r}')\psi(\boldsymbol{r}')d^3\boldsymbol{r}'$$

where $G(\boldsymbol{r} - \boldsymbol{r}')$ is a Green's function and equals

$$G(\boldsymbol{r}-\boldsymbol{r}') = \frac{1}{(2\pi)^3}\int \frac{e^{iq(\boldsymbol{r}-\boldsymbol{r}')}}{k^2 - q^2}d^3q$$

This integral breaks down to

$$G(\boldsymbol{r}-\boldsymbol{r}') = \frac{-1}{4\pi^2 i|\boldsymbol{r}-\boldsymbol{r}'|}\int \frac{qe^{iq|\boldsymbol{r}-\boldsymbol{r}'|}}{q^2 - k^2}dq$$

You can solve this integral in terms of incoming and/or outgoing waves, and the Green's function takes the following form:

$$G(\boldsymbol{r}-\boldsymbol{r}') = \frac{-1}{4\pi}\frac{e^{ik|\boldsymbol{r}-\boldsymbol{r}'|}}{|\boldsymbol{r}-\boldsymbol{r}'|}$$

So here's a messy version of the Schrödinger equation that you use to find the scattering amplitude. The first term is the homogeneous solution, and the second term is the particular solution with Green's function filled in:

$$\psi(\boldsymbol{r}) = Ae^{i\boldsymbol{k}_0\cdot\boldsymbol{r}} + \frac{\mu}{2\pi\hbar^2}\int \frac{e^{ik|\boldsymbol{r}-\boldsymbol{r}'|}}{|\boldsymbol{r}-\boldsymbol{r}'|}V(\boldsymbol{r}')\psi(\boldsymbol{r}')d^3\boldsymbol{r}'$$

To solve this equation, you use the *Born approximation,* a series expansion that provides you with terms that match the actual solution successively more closely. The example and practice problems show you how.

Q. Find the zeroth-order Born approximation — that is, find the zeroth-order term in the following equation (the term before any corrections are added in).

$$\psi(\mathbf{r}) = Ae^{i\mathbf{k}_0\cdot\mathbf{r}} + \frac{\mu}{2\pi\hbar^2}\int \frac{e^{ik|\mathbf{r}-\mathbf{r}'|}}{|\mathbf{r}-\mathbf{r}'|}V(\mathbf{r}')\psi(\mathbf{r}')d^3\mathbf{r}'$$

A. **$_0(\mathbf{r})$ = $_{\text{inc}}(\mathbf{r})$.** The problem here is to solve

$$\psi(\mathbf{r}) = \phi_{\text{inc}} + \frac{\mu}{2\pi\hbar^2}\int \frac{e^{ik|\mathbf{r}-\mathbf{r}'|}}{|\mathbf{r}-\mathbf{r}'|}V(\mathbf{r}')\psi(\mathbf{r}')d^3\mathbf{r}'$$

where $\phi_{\text{inc}} = Ae^{i\mathbf{k}_0\cdot\mathbf{r}}$. Break the problem into an initial term and a correction term:

$$\psi(\mathbf{r}) = \phi_{\text{inc}} + \frac{\mu}{2\pi\hbar^2}\int \frac{e^{ik|\mathbf{r}-\mathbf{r}'|}}{|\mathbf{r}-\mathbf{r}'|}V(\mathbf{r}')\psi(\mathbf{r}')d^3\mathbf{r}'$$
$$= \psi_0(\mathbf{r}) + \psi_{\text{correction}}(\mathbf{r})$$

The second term is the correction term, so the zeroth-order term in the approximation is just the first term:

$$\psi_0(\mathbf{r}) = \phi_{\text{inc}}(\mathbf{r})$$

9. Find the first-order Born approximation. You get the first-order term by substituting $v'_{2c} = v_{2c}$ into

$$\frac{1}{2}m_1v_{1c}^2 + \frac{1}{2}m_2v_{2c}^2 = \frac{1}{2}m_1v_{1c}'^2 + \frac{1}{2}m_2v_{2c}'^2$$

$$\psi_2(\mathbf{r}) = \phi_{\text{inc}} - \frac{\mu}{2\pi\hbar^2}\int \frac{e^{ik|\mathbf{r}-\mathbf{r}_2|}}{|\mathbf{r}-\mathbf{r}_2|}V(\mathbf{r}_2)\psi_1(\mathbf{r}_2)d^3\mathbf{r}_2$$

Solve It

10. Find the second-order Born approximation by plugging the first-order Born approximation into

$$\psi(\mathbf{r}) = \phi_{\text{inc}} + \frac{\mu}{2\pi\hbar^2}\int \frac{e^{ik|\mathbf{r}-\mathbf{r}'|}}{|\mathbf{r}-\mathbf{r}'|}V(\mathbf{r}')\psi(\mathbf{r}')d^3\mathbf{r}'$$

Solve It

Putting the Born Approximation to the Test

So is the Born approximation your magic pill? Can it solve even the odd potentials you may encounter? The good news is you can push the Born approximation to the limit and see exactly what it can do to solve various types of scattering problems. For example, for weak, spherically symmetric potentials, the differential cross section looks like the following:

$$\frac{d\sigma(\phi, \theta)}{d\Omega} = \frac{-4\mu^2}{q^2\hbar^4}\left|\int_0^\infty r'\,V(r')\sin(qr')dr'\right|^2$$

For example, you may find the differential cross section for two electrically charged particles of charge Z_1e and Z_2e, where the potential looks like this:

$$V(r) = \frac{Z_1Z_2e^2}{r}$$

The following text explores how to use the Born approximation with some problems.

Q. Find the differential cross section in integral form for two electrically charged particles of charge Z_1e and Z_2e, where the potential looks like this:

$$V(r) = \frac{Z_1Z_2e^2}{r}$$

A. $$\frac{d\sigma(\phi, \theta)}{d\Omega} = \frac{-4Z_1^2Z_2^2e^4\mu^2}{q^2\hbar^4}\left|\int_0^\infty \sin(qr')dr'\right|^2$$

The differential cross section looks like this for a weak, spherically symmetrical potential:

$$\frac{d\sigma(\phi, \theta)}{d\Omega} = \frac{-4\mu^2}{q^2\hbar^4}\left|\int_0^\infty r'\,V(r')\sin(qr')dr'\right|^2$$

Therefore, here's what the differential cross section looks like using the first Born approximation in integral form:

$$\frac{d\sigma(\phi, \theta)}{d\Omega} = \frac{-4Z_1^2Z_2^2e^4\mu^2}{q^2\hbar^4}\left|\int_0^\infty \sin(qr')dr'\right|^2$$

11. Find the differential cross section by solving this integral for two electrically charged particles of charge Z_1e and Z_2e:

$$\frac{d\sigma(\phi,\theta)}{d\Omega}=\frac{-4Z_1^2Z_2^2e^4\mu^2}{q^2\hbar^4}\left|\int_0^\infty \sin(qr')\,dr'\right|^2$$

Solve It

12. You smash an alpha particle, $Z_1 = 4$, against a lead nucleus, $Z_2 = 82$. If the scattering angle in the lab frame is 58°, what is it in the center-of-mass frame and what is the differential cross section?

Solve It

Answers to Problems on Scattering Theory

The following are the answers to the practice questions I present earlier in this chapter.

1 How do you relate the total cross section, σ, to the differential cross section, $d\sigma(\phi, \theta)/d\Omega$, in terms of Ω? **The answer is**

$$\sigma = \int \frac{d\sigma(\phi, \theta)\, d\Omega}{d\Omega}$$

Start with the differential cross section:

$$\frac{d\sigma(\phi, \theta)}{d\Omega}$$

The total cross section, σ, is the sum of the differential cross section over all angles, so you integrate it, giving you the following:

$$\sigma = \int \frac{d\sigma(\phi, \theta)\, d\Omega}{d\Omega}$$

2 How do you relate the total cross section, σ, to the differential cross section, $d\sigma(\phi, \theta)/d\Omega$, in terms of θ and ϕ? **The answer is**

$$\sigma = \int \frac{d\sigma(\phi, \theta)\, d\Omega}{d\Omega} = \int_0^{\pi} \sin\theta\, d\theta \int_0^{2\pi} \frac{d\sigma(\phi, \theta)\, d\phi}{d\Omega}$$

The total cross section, σ, is the sum of the differential cross section over all angles, so it's equal to

$$\sigma = \int \frac{d\sigma(\phi, \theta)\, d\Omega}{d\Omega}$$

And because $d\Omega = \sin\theta\, d\theta\, d\phi$, the total cross section, σ, is equal to

$$\sigma = \int \sin\theta\, d\theta \int \frac{d\sigma(\phi, \theta)\, d\phi}{d\Omega}$$

Put in the limits of integration for a sphere to get the final answer:

$$\sigma = \int_0^{\pi} \sin\theta\, d\theta \int_0^{2\pi} \frac{d\sigma(\phi, \theta)\, d\phi}{d\Omega}$$

3 Start with this relation:

$$\tan\theta_1 = \frac{\sin\theta}{\cos\theta + v_{cm}/v'_{1c}}$$

Relate θ_1 and θ without involving the velocities; use only the masses of the two particles, given that the collision is elastic (that is, kinetic energy is conserved) and $\boldsymbol{v}_{cm} = \frac{m_1}{m_1 + m_2}\, \boldsymbol{v}_{1lab}$ and $\boldsymbol{v}_{1c} = \frac{m_2}{m_1 + m_2}\, \boldsymbol{v}_{1lab}$.

The answer is

$$\tan\theta_1 = \frac{\sin\theta}{\cos\theta + m_1/m_2}$$

Start with the givens in the problem, the incident velocities in the lab and center-of-mass frames:

- $v_{cm} = \frac{m_1}{m_1 + m_2} v_{1lab}$
- $v_{1c} = \frac{m_2}{m_1 + m_2} v_{1lab}$

Because the center of mass is stationary in the center-of-mass frame, the total momentum before and after the collision is zero in that frame, so

$$m_1 v_{1c} - m_2 v_{2c} = 0$$

Solve for v_{2c}:

$$v_{2c} = \frac{m_1}{m_2} v_{1c}$$

The components of momentum are also conserved, so $m_1 v'_{1c} \cos\theta - m_2 v'_{2c} \cos\theta = 0$, which means that

$$v'_{2c} = \frac{m_1}{m_2} v'_{1c}$$

Because you're assuming all collisions are elastic, kinetic energy is conserved in addition to momentum, so the following is true:

$$\frac{1}{2} m_1 v_{1c}^2 + \frac{1}{2} m_2 v_{2c}^2 = \frac{1}{2} m_1 v_{1c}'^2 + \frac{1}{2} m_2 v_{2c}'^2$$

So you can relate the center-of-mass and lab velocities like this:

$$v'_{1c} = v_{1c} \text{ and } v'_{2c} = v_{2c}$$

This gives you

$$v'_{1c} = v_{1c} = \frac{m_2}{m_1 + m_2} v_{1lab}$$

Dividing v_{cm} by v'_{1c} gives you the following:

$$\frac{v_{cm}}{v'_{1c}} = \frac{m_1}{m_2}$$

From the example problem in this section, you know that

$$\tan\theta_1 = \frac{\sin\theta}{\cos\theta + v_{cm}/v'_{1c}}$$

Therefore, you can substitute for v'_{cm}/v'_{1c}. Doing so gives you

$$\tan\theta_1 = \frac{\sin\theta}{\cos\theta + m_1/m_2}$$

4 Relate θ_2 to θ. Given that $\tan\theta_2 = \cot(\theta/2)$, show that

$$\theta_2 = \frac{\pi - \theta}{2}$$

You end with

$$\boldsymbol{\theta_2 = \frac{\pi - \theta}{2}}$$

Start with the given:

$$\tan\theta_2 = \cot(\theta/2)$$

Plug in that tan = sin/cos and cot = 1/tan (or cos/sin) to get the following:

$$\frac{\sin\theta_2}{\cos\theta_2} = \frac{\cos(\theta/2)}{\sin(\theta/2)}$$

Now convert from sine to cosine and cosine to sine this way:

$$\frac{\sin\theta_2}{\cos\theta_2} = \frac{\sin(\pi/2 - \theta/2)}{\cos(\pi/2 - \theta/2)}$$

Now convert back to tangents:

$$\tan\theta_2 = \tan(\pi/2 - \theta/2)$$

Take the inverse tangent to get the final answer:

$$\theta_2 = \frac{\pi - \theta}{2}$$

5 Start with the following relation:

$$\left.\frac{d\sigma(\phi,\theta)}{d\Omega_1}\right|_{lab} = \left.\frac{d\sigma(\phi,\theta)}{d\Omega}\right|_{cm} \frac{d(\cos\theta)}{d(\cos\theta_1)}$$

Convert it to this, adding the masses:

$$\left.\frac{d\sigma(\phi,\theta)}{d\Omega_1}\right|_{lab} = \frac{\left(1 + \frac{m_1^2}{m_2^2} + 2\cos\theta\left(\frac{m_1}{m_2}\right)\right)^{3/2}}{1 + \cos\theta\left(\frac{m_1}{m_2}\right)} \left.\frac{d\sigma(\phi,\theta)}{d\Omega}\right|_{cm}$$

End with

$$\left.\frac{d\sigma(\phi,\theta)}{d\Omega_1}\right|_{\text{lab}} = \frac{\left(1+\frac{m_1^2}{m_2^2}+2\cos\theta\left(\frac{m_1}{m_2}\right)\right)^{3/2}}{1+\cos\theta\left(\frac{m_1}{m_2}\right)}\left.\frac{d\sigma(\phi,\theta)}{d\Omega}\right|_{\text{cm}}$$

Start with the given:

$$\left.\frac{d\sigma(\phi,\theta)}{d\Omega_1}\right|_{\text{lab}} = \left.\frac{d\sigma(\phi,\theta)}{d\Omega}\right|_{\text{cm}}\frac{d(\cos\theta)}{d(\cos\theta_1)}$$

In problem 3, you relate θ_1 to θ like this:

$$\tan\theta_1 = \frac{\sin\theta}{\cos\theta+\frac{m_1}{m_2}}$$

If you want to keep everything in terms of sines and cosines, you can use the following relation:

$$\cos\theta_1 = \frac{1}{\left(\tan^2\theta_1+1\right)^{1/2}}$$

Here's what you get by using your knowledge of trigonometry:

$$\cos\theta_1 = \frac{\cos\theta+\frac{m_1}{m_2}}{\left(1+\frac{m_1^2}{m_2^2}+2\cos\theta\left(\frac{m_1}{m_2}\right)\right)^{1/2}}$$

Therefore, taking the derivative with respect to cos θ gives you the following:

$$\frac{d(\cos\theta_1)}{d(\cos\theta)} = \frac{1+\cos\theta\left(\frac{m_1}{m_2}\right)}{\left(1+\frac{m_1^2}{m_2^2}+2\cos\theta\left(\frac{m_1}{m_2}\right)\right)^{3/2}}$$

And plugging this into the earlier equation, you get

$$\left.\frac{d\sigma(\phi,\theta)}{d\Omega_1}\right|_{\text{lab}} = \frac{\left(1+\frac{m_1^2}{m_2^2}+2\cos\theta\left(\frac{m_1}{m_2}\right)\right)^{3/2}}{1+\cos\theta\left(\frac{m_1}{m_2}\right)}\left.\frac{d\sigma(\phi,\theta)}{d\Omega}\right|_{\text{cm}}$$

6 Start with two particles of equal mass colliding in the lab frame, where one particle starts at rest. Show that the two particles end up traveling at right angles with respect to each other in the lab frame. **The answer is $_2 + {}_1 = /2$.**

Start with the equation for two particles colliding in the lab frame (see problem 3 for the derivation):

$$\cos\theta_1 = \frac{\cos\theta + \dfrac{m_1}{m_2}}{\left(1+\dfrac{m_1^2}{m_2^2}+2\cos\theta\left(\dfrac{m_1}{m_2}\right)\right)^{1/2}}$$

The masses are equal. If $m_1 = m_2$, then you get

$\tan(\theta_1) = \tan(\%)$

Therefore, $\theta_1 = \%$.

You've also seen that the following is true (see problem 5 for the calculations):

$$\left.\frac{d\sigma(\phi,\theta)}{d\Omega_1}\right|_{lab} = \frac{\left(1+\dfrac{m_1^2}{m_2^2}+2\cos\theta\left(\dfrac{m_1}{m_2}\right)\right)^{3/2}}{1+\cos\theta\left(\dfrac{m_1}{m_2}\right)}\left.\frac{d\sigma(\phi,\theta)}{d\Omega}\right|_{cm}$$

And by factoring, this becomes

$$\left.\frac{d\sigma(\phi,\theta)}{d\Omega_1}\right|_{lab} = 4\cos(\theta/2)\left.\frac{d\sigma(\phi,\theta)}{d\Omega}\right|_{cm}$$

Note that

$\tan(\theta_2) = \cot(\%)$

$= \tan(\% - \%)$

Therefore, $\theta_2 = \% - \%$. Because $\theta_1 = \%$ and $\theta_2 = \% - \%$, you know that $\theta_2 = \% - \% = \% - \theta_1$. In other words

$\theta_2 + \theta_1 = \%$

So the particles end up at right angles in the lab frame.

7 Find the wave function of an incident particle. Assume that the scattering potential V(*r*) has a very finite range and find the wave function outside that range. **The answer is $\phi_{inc}(\mathbf{r}) = \mathbf{A}e^{i\mathbf{k}_0\cdot\mathbf{r}}$, where $k_0^2 = \dfrac{2\mu\mathbf{E}_0}{\hbar^2}$.**

Assume that the scattering potential V(***r***) has a very finite range. Outside that range, the wave functions involved act like free particles. The incident particle's wave function, outside the limit of V(*r*), is given by the following equation, because V(***r***) is zero:

$$\nabla^2\phi_{\text{inc}}(\boldsymbol{r})+k_0\phi_{\text{inc}}(\boldsymbol{r})=0$$

where $k_0^2=\frac{2\mu E_0}{\hbar^2}$ and E_0 is the energy of the incident particle. Solving this equation gives you the following:

$$\phi_{\text{inc}}(\boldsymbol{r})=Ae^{i\boldsymbol{k}_0\cdot\boldsymbol{r}}$$

where A is a normalization factor and $\boldsymbol{k}_0\cdot\boldsymbol{r}$ is the dot product between the incident wave's wave vector and ***r***.

8 The wave function of a scattered particle looks like this:

$$\phi_{\text{sc}}(\boldsymbol{r})=Af(\phi,\theta)\frac{e^{ik\cdot r}}{r}$$

where f(ϕ, θ) is the scattering amplitude, A is a dimensionless normalization factor, and $\left|\boldsymbol{k}^2\right|=\frac{2\mu E}{\hbar^2}$. Find the dimensions of f(ϕ, θ). **The dimensions of f(,) are length.**

The scattered wave function looks like this:

$$\phi_{\text{sc}}(r)=Af(\phi,\theta)\frac{e^{ik\cdot r}}{r}$$

Here, A is a normalization factor and $\left|\boldsymbol{k}^2\right|=\frac{2\mu E}{\hbar^2}$, where E is the energy of the scattered particle. The wave function must be normalized to 1, so $\phi_{\text{sc}}(r)$ must be dimensionless. The constant A is dimensionless, so f(ϕ, θ) must have units of length to cancel out the units of length from *r* in the denominator of $\phi_{\text{sc}}(r)$.

9 Find the first-order Born approximation. You get the first-order term by substituting $v'_{2c}=v_{2c}$ into

$$\frac{1}{2}m_1v_{1c}^2+\frac{1}{2}m_2v_{2c}^2=\frac{1}{2}m_1v_{1c}'^2+\frac{1}{2}m_2v_{2c}'^2$$

$$\psi_2(\boldsymbol{r})=\phi_{\text{inc}}-\frac{\mu}{2\pi\hbar^2}\int\frac{e^{ik|\boldsymbol{r}-\boldsymbol{r}_2|}}{|\boldsymbol{r}-\boldsymbol{r}_2|}V(\boldsymbol{r}_2)\psi_1(\boldsymbol{r}_2)d^3\boldsymbol{r}_2$$

The answer is

$$\boldsymbol{\psi_1(r)=\phi_{\text{inc}}-\frac{\mu}{2\pi\hbar^2}\int\frac{e^{ik|r-r_1|}}{|r-r_1|}V(r_1)\phi_{\text{inc}}(r_1)d^3r_1}$$

Substitute the zeroth-order term, $\psi_0(\boldsymbol{r})$, into the following equation:

$$\psi(\boldsymbol{r})=\phi_{\text{inc}}+\frac{\mu}{2\pi\hbar^2}\int\frac{e^{ik|\boldsymbol{r}-\boldsymbol{r}'|}}{|\boldsymbol{r}-\boldsymbol{r}'|}V(\boldsymbol{r}')\psi(\boldsymbol{r}')d^3\boldsymbol{r}'$$

This gives you first-order term:

$$\psi_1(\mathbf{r}) = \phi_{\text{inc}} + \frac{\mu}{2\pi\hbar^2}\int \frac{e^{ik|\mathbf{r}-\mathbf{r}_1|}}{|\mathbf{r}-\mathbf{r}_1|} V(\mathbf{r}_1)\psi_0(\mathbf{r}_1)d^3\mathbf{r}_1$$

Using $\psi_0(\mathbf{r}) = \phi_{\text{inc}}(\mathbf{r})$, you get the following answer:

$$\psi_1(\mathbf{r}) = \phi_{\text{inc}} - \frac{\mu}{2\pi\hbar^2}\int \frac{e^{ik|\mathbf{r}-\mathbf{r}_1|}}{|\mathbf{r}-\mathbf{r}_1|} V(\mathbf{r}_1)\phi_{\text{inc}}(\mathbf{r}_1)d^3\mathbf{r}_1$$

10 Find the second-order Born approximation by plugging the first-order Born approximation into

$$\psi(\mathbf{r}) = \phi_{\text{inc}} + \frac{\mu}{2\pi\hbar^2}\int \frac{e^{ik|\mathbf{r}-\mathbf{r}'|}}{|\mathbf{r}-\mathbf{r}'|} V(\mathbf{r}')\psi(\mathbf{r}')d^3\mathbf{r}'$$

The answer is

$$\boldsymbol{\psi_2(\mathbf{r}) = \phi_{\text{inc}} - \frac{\mu}{2\pi\hbar^2}\int \frac{e^{ik|\mathbf{r}-\mathbf{r}_2|}}{|\mathbf{r}-\mathbf{r}_2|} V(\mathbf{r}_2)\phi_{\text{inc}}(\mathbf{r}_2)d^3\mathbf{r}_2}$$

$$\boldsymbol{+\frac{\mu^2}{4\pi^2\hbar^4}\int \frac{e^{ik|\mathbf{r}-\mathbf{r}_2|}}{|\mathbf{r}-\mathbf{r}_2|} V(\mathbf{r}_2)\psi_1(\mathbf{r}_2)d^3\mathbf{r}_2 \int \frac{e^{ik|\mathbf{r}_2-\mathbf{r}_1|}}{|\mathbf{r}_2-\mathbf{r}_1|} V(\mathbf{r}_1)\phi_{\text{inc}}(\mathbf{r}_1)d^3\mathbf{r}_1}$$

You get the second-order term by substituting the following into the given equation:

- $\psi_1(\mathbf{r}) = \phi_{\text{inc}} + \frac{\mu}{2\pi\hbar^2}\int \frac{e^{ik|\mathbf{r}-\mathbf{r}_1|}}{|\mathbf{r}-\mathbf{r}_1|} V(\mathbf{r}_1)\psi_0(\mathbf{r}_1)d^3\mathbf{r}_1$
- $\psi(\mathbf{r}) = \phi_{\text{inc}} + \frac{\mu}{2\pi\hbar^2}\int \frac{e^{ik|\mathbf{r}-\mathbf{r}'|}}{|\mathbf{r}-\mathbf{r}'|} V(\mathbf{r}')\psi(\mathbf{r}')d^3\mathbf{r}'$

Here's what the substitution gives you:

$$\psi_2(\mathbf{r}) = \phi_{\text{inc}} - \frac{\mu}{2\pi\hbar^2}\int \frac{e^{ik|\mathbf{r}-\mathbf{r}_2|}}{|\mathbf{r}-\mathbf{r}_2|} V(\mathbf{r}_2)\psi_1(\mathbf{r}_2)d^3\mathbf{r}_2$$

So you get

$$\psi_2(\mathbf{r}) = \phi_{\text{inc}} - \frac{\mu}{2\pi\hbar^2}\int \frac{e^{ik|\mathbf{r}-\mathbf{r}_2|}}{|\mathbf{r}-\mathbf{r}_2|} V(\mathbf{r}_2)\phi_{\text{inc}}(\mathbf{r}_2)d^3\mathbf{r}_2$$

$$+\frac{\mu^2}{4\pi^2\hbar^4}\int \frac{e^{ik|\mathbf{r}-\mathbf{r}_2|}}{|\mathbf{r}-\mathbf{r}_2|} V(\mathbf{r}_2)\psi_1(\mathbf{r}_2)d^3\mathbf{r}_2 \int \frac{e^{ik|\mathbf{r}_2-\mathbf{r}_1|}}{|\mathbf{r}_2-\mathbf{r}_1|} V(\mathbf{r}_1)\phi_{\text{inc}}(\mathbf{r}_1)d^3\mathbf{r}_1$$

11 Find the differential cross section by solving this integral for two electrically charged particles of charge Z_1e and Z_2e:

$$\frac{d\sigma(\phi,\theta)}{d\Omega} = \frac{-4Z_1^2Z_2^2e^4\mu^2}{q^2\hbar^4}\left|\int_0^\infty \sin(qr')dr'\right|^2$$

The answer is

$$\boldsymbol{\frac{d\sigma(\phi,\theta)}{d\Omega} = \frac{Z_1^2Z_2^2e^4}{16E^2}\sin^{-4}\left(\frac{\theta}{2}\right)}$$

where E is the kinetic energy of the incoming particle:

$$\mathbf{E} = \frac{\hbar^2 \boldsymbol{k}^2}{2\mu}$$

Start with the integral you want to solve for two electrically charged particles of charge $Z_1 e$ and $Z_2 e$:

$$\frac{d\sigma(\phi,\theta)}{d\Omega} = \frac{4Z_1^2 Z_2^2 e^4 \mu^2}{q^2 \hbar^4} \left| \int_0^\infty \sin(qr')\,dr' \right|^2$$

Take the integral:

$$\int_0^\infty \sin(qr')\,dr' = \frac{1}{q}$$

Plug in $(1/q)^2$ into the original equation:

$$\frac{d\sigma(\phi,\theta)}{d\Omega} = \frac{4Z_1^2 Z_2^2 e^4 \mu^2}{q^4 \hbar^4}$$

So recalling that $q = 2k \sin(\theta/2)$, you know that

$$\frac{d\sigma(\phi,\theta)}{d\Omega} = \frac{Z_1^2 Z_2^2 e^4}{16E^2} \sin^{-4}\left(\frac{\theta}{2}\right)$$

where E is the kinetic energy of the incoming particle:

$$E = \frac{\hbar^2 k^2}{2\mu}$$

12 You smash an alpha particle, $Z_1 = 4$, against a lead nucleus, $Z_2 = 82$. If the scattering angle in the lab frame is 58°, what is it in the center-of-mass frame and what is the differential cross section? **The answer is $= 59°$ and $\frac{d\sigma(\phi,\theta)}{d\Omega} = 3.1 \times 10^{-29}\,\mathbf{m}^2$.**

The ratio of the particle's mass, m_1/m_2, is 0.02, so the scattering angle in the center of mass frame, θ, is

$$\tan\theta_{lab} = \frac{\sin\theta}{\cos\theta + 0.02}$$

Here, $\theta_{lab} = 58°$. Solving for θ gives you $\theta = 59°$.

Here's the differential cross section (see problem 11 for the calculations):

$$\frac{d\sigma(\phi,\theta)}{d\Omega} = \frac{4Z_1^2 Z_2^2 e^4 \mu^2}{q^4 \hbar^4} = \frac{Z_1^2 Z_2^2 e^4}{16E^2} \sin^{-4}\left(\frac{\theta}{2}\right)$$

Substitute the numbers of the incident alpha particle's energy (8 MeV) to get the final answer:

$$\frac{d\sigma(\phi,\theta)}{d\Omega} = 3.1 \times 10^{-29}\,\text{m}^2$$

Part V
The Part of Tens

"This is my old physics teacher, Mr. Wendt, his wife Doris, and their two children, Bra and Ket."

In this part . . .

The Part of the Tens is traditional in all *For Dummies* books, and here you see ten tips for solving quantum physics problem, ten famous problems that quantum physics answered, and ten pitfalls to watch out for when solving quantum physics problems.

Chapter 12

Ten Tips to Make Solving Quantum Physics Problems Easier

In This Chapter

- Rewriting equations into simpler terms
- Working with operators
- Using tables
- Breaking equations into parts

Whether you've recently started solving quantum physics problems or you've been working through them for a while, one thing is certain: Quantum physics offers lots of places where you can get stuck or take a wrong turn. The good news: Although quantum physics is no walk in the park, I provide ten tips to make your journey a bit easier. Take a look at the tips here — they may save you a lot of time and headaches.

Normalize Your Wave Functions

You need to normalize a wave function before you can work with it in a general way. *Normalizing* the wave function means that the total probability of the particle's appearing somewhere in space is 1. Make sure you normalize your wave functions to 1 as in the following example:

$$1=\int_{-\infty}^{\infty}\left|\psi(x)\right|^{2}dx$$

Check out Chapter 2 for the ins and outs of normalizing wave functions.

Use Eigenvalues

Eigenvalues are the values you get when you apply an operator to an eigenstate of that operator. You may apply an operator to a state vector and get a value, but that value isn't an *eigenvalue* if the state vector changes:

$$A\left|\psi\right\rangle=a\left|\chi\right\rangle$$

Here's the way eigenvalues should behave — note that the state vector remains unchanged:

$$A|\psi\rangle = a|\psi\rangle$$

Chapter 1 delves deeper into how eigenvalues work.

Meet the Boundary Conditions for Wave Functions

In quantum physics, boundary conditions don't have anything to do with international relations; boundary conditions give you the information you need to solve a problem. You solve the problem up to a set of undetermined constants and then use the boundary conditions to finish. For example, consider a wave-function problem. Take a look at this square well:

$$V(x) = \begin{cases} \infty & x < 0 \\ 0 & 0 \le x \le a \\ \infty & x > a \end{cases}$$

These boundary conditions mean that

- $\psi(0) = 0$
- $\psi(a) = 0$

Using the boundary conditions, you can solve the Schrödinger equation to get the wave function:

$$\psi(a) = A \sin(ka) = 0$$

Refer to Chapter 2 for more information on boundary conditions for wave functions.

Meet the Boundary Conditions for Energy Levels

Boundary conditions can also help you find the energy levels. You solve for the energy levels up to arbitrary constants and then apply the boundary conditions to finish the solution. Consider the following square well:

$$V(x) = \begin{cases} \infty & x < 0 \\ 0 & 0 \le x \le a \\ \infty & x > a \end{cases}$$

Here's what the wave function looks like:

$$\psi(a) = A \sin(ka) = 0$$

Solving for the energy levels isn't hard; the boundary conditions mean that

$$ka = n\pi \qquad n = 1, 2, 3, \ldots$$

Now solve for k, the wave number:

$$k = \frac{n\pi}{a} \quad n = 1, 2, 3, \ldots$$

And that means you can constrain the energy this way:

$$\frac{2mE}{\hbar^2} = \frac{n^2\pi^2}{a^2} \quad n = 1, 2, 3, \ldots$$

Solving for E gives you the final energy-levels formula:

$$E = \frac{n^2\hbar^2\pi^2}{2ma^2} \quad n = 1, 2, 3, \ldots$$

Check out Chapter 2 for in-depth material on boundary conditions for energy levels.

Use Lowering Operators to Find the Ground State

The trick to finding the wave function for a system is often finding the ground state using the *lowering operator,* which basically gives you the energy state one lower than the current one. The *ground state* of a particle is its lowest energy level. Applying the lowering operator to the first excited state should give you the ground state so that you have an actual result to work with. The following equation can help:

$$a|0\rangle = 0$$

Applying the $\langle x|$ bra gives you the following:

$$\langle x|a|0\rangle = 0$$

In the case of a harmonic oscillator, substitute for a using its position representation:

$$\langle x|a|0\rangle = \frac{1}{x_0\sqrt{2}}\left\langle x\left|X + x_0^2\frac{d}{dx}\right|0\right\rangle = 0$$

Then you use $\langle x|0\rangle = \psi_0(x)$ to get the following:

$$\frac{1}{x_0\sqrt{2}}\left(x\,\psi_0(x) + x_0^2\frac{d\psi_0(x)}{dx}\right) = 0$$

Multiplying both sides by $x_0\sqrt{2}$ gives you

$$x\,\psi_0(x) + x_0^2\frac{d\psi_0(x)}{dx} = 0$$

Rearrange the equation:

$$\frac{d\psi_0(x)}{dx} = -\frac{x}{x_0^2}\psi_0(x)$$

Solve this, and you get the following answer for the ground state:

$$\psi_0(x) = A\exp\left(\frac{-x^2}{2x_0^2}\right)$$

Use Raising Operators to Find the Excited States

If you use a lowering operator to solve for the ground state (see the preceding section, "Use Lowering Operators to Find the Ground State"), you use raising operators to solve for the excited states. Although you probably picture a 3-year-old running around in an excited state on Halloween night, an *excited state* in quantum physics is a higher-energy state. The *raising operator* is the operator that gets you to a higher state. For example, the first excited state is

$$\psi_1(x) = \langle x|1\rangle$$

You can use the raising operator, $a^\dagger$, on the ground state:

$$|1\rangle = a^\dagger|0\rangle$$

So in the position representation

$$\psi_1(x) = \langle x|a^\dagger|0\rangle$$

For example, for a harmonic oscillator, the raising operator is

$$a^\dagger = \frac{1}{x_0\sqrt{2}}\left(X - x_0^2\frac{d}{dx}\right)$$

So plug the value for $a^\dagger$ into the $\psi_1(x)$ equation:

$$\langle x|a^\dagger|0\rangle = \frac{1}{x_0\sqrt{2}}\left\langle x\left|\left(X - x_0^2\frac{d}{dx}\right)\right|0\right\rangle$$
$$= \frac{1}{x_0\sqrt{2}}\left(X - x_0^2\frac{d}{dx}\right)\langle x|0\rangle$$

And because $\psi_0(x) = \langle x|0\rangle$, you can make the following substitution:

$$\psi_1(x) = \frac{1}{x_0\sqrt{2}}\left(X - x_0^2\frac{d}{dx}\right)\psi_0(x)$$

This equals the following:

$$\psi_1(x)=\frac{1}{x_0\sqrt{2}}\left(x-x_0^2\left(-\frac{x}{x_0^2}\right)\right)\psi_0(x)$$

Rationalize the fraction and simplify the equation:

$$\psi_1(x)=\frac{\sqrt{2}}{x_0}x\,\psi_0(x)$$

You know the ground state, so you can plug in the equation for $\psi_0(x)$. Because $\psi_0(x)=\frac{1}{\pi^{1/4}x_0^{1/2}}\exp\left(\frac{-x^2}{2x_0^2}\right)$, you get the following answer for the first excited state:

$$\psi_1(x)=\frac{\sqrt{2}}{\pi^{1/4}x_0^{3/2}}x\exp\left(\frac{-x^2}{2x_0^2}\right)$$

Use Tables of Functions

In quantum physics, you usually have two ways to determine the form of a function: Use a generator equation or use a table of functions. Using a table of functions is usually a safer bet. For example, here's a well-known formula for the wave function of a harmonic oscillator:

$$\psi_n(x)=\frac{1}{\pi^{1/4}\left(2^n n!x_0\right)^{1/2}}\mathrm{H}_n\left(\frac{x}{x_0}\right)\exp\left(\frac{-x^2}{2x_0^2}\right)$$

where $x_0=\left(\hbar/(m\omega)\right)^{1/2}$.

$\mathrm{H}_n(x)$ is the nth Hermite polynomial, which is defined this way:

$$\mathrm{H}_n(x)=(-1)^n\exp\left(x^2\right)\frac{d^n}{dx^n}\left(-x^2\right)$$

Many people try to derive complex functions like Hermite polynomials themselves. But other people have already done the work for you, so why take chances? You can get polynomials like these from tables of functions, like this:

- $\mathrm{H}_0(x) = 1$
- $\mathrm{H}_1(x) = 2x$
- $\mathrm{H}_2(x) = 4x^2 - 2$
- $\mathrm{H}_3(x) = 8x^3 - 12x$
- $\mathrm{H}_4(x) = 16x^4 - 48x^2 + 12$
- $\mathrm{H}_5(x) = 32x^5 - 160x^3 + 120x$

Decouple the Schrödinger Equation

When you can, break the Schrödinger equation into parts that make it easier to solve. If you have four particles that interact only weakly, for example, ignore that weak interaction to decouple the Schrödinger equation into four independent parts. Multiple simple Schrödinger equations are always easier to solve than one giant complex one.

$$E\psi_{n_1,n_2,n_3,n_4}(\boldsymbol{r}_1,\boldsymbol{r}_2,\boldsymbol{r}_3,\boldsymbol{r}_4)=\sum_{i=1}^{4}-\frac{\hbar^2}{2m_i}\frac{d^2}{dx_i^2}\psi_{n_1,n_2,n_3,n_4}(\boldsymbol{r}_1,\boldsymbol{r}_2,\boldsymbol{r}_3,\boldsymbol{r}_4)$$

Use Two Schrödinger Equations for Hydrogen

The Schrödinger equation lets you solve for a particle's wave function. But the Schrödinger equation for hydrogen can be intertwined for the electron and the proton, so it's best if you decouple the two.

Recast the Schrödinger equation to use **R** and $\boldsymbol{r}$ instead of $\boldsymbol{r}_e$ (the position of the electron) and $\boldsymbol{r}_p$ (the position of the proton), where

$$\mathbf{R}=\frac{m_e\boldsymbol{r}_e+m_p\boldsymbol{r}_p}{m_e+m_p}$$

is the center of mass and $\boldsymbol{r}$ is the difference between $\boldsymbol{r}_e$ and $\boldsymbol{r}_p$:

$$\boldsymbol{r}=\boldsymbol{r}_e-\boldsymbol{r}_p$$

This change decouples the Schrödinger equation to the following two equations:

- $-\frac{\hbar^2}{2M}\nabla_R^2\psi(\mathbf{R})=E_R\psi(\mathbf{R})$
- $-\frac{\hbar^2}{2m}\nabla_r^2\psi(\boldsymbol{r})-\frac{e^2}{|\boldsymbol{r}|}\psi(\boldsymbol{r})=E_r\psi(\boldsymbol{r})$

This is much easier to solve than the Schrödinger equation in terms of $\boldsymbol{r}_e$ and $\boldsymbol{r}_p$.

Take the Math One Step at a Time

When you're solving quantum physics problems, some people tend to get overwhelmed. Take a deep breath and don't get lost in the math. Yes, there's a lot of math in quantum physics, but you can master it. Take it one step at a time, show your work, and check everything twice. With time, what seems impossible will become just another skill. Good luck with it!

Chapter 13

Ten Famous Solved Quantum Physics Problems

In This Chapter

- Locating and describing particles
- Getting insight into atomic behavior
- Seeing the light on photons

Quantum physics is famous for the problems it's solved. Some, like spin, take special equipment to see; others, like the spectrum of hydrogen in stars, don't. There are only a limited number of systems that quantum physics has solved through and through, so when someone comes up with a breakthrough that really explains a system, it's big news.

This chapter provides you with a list of some of quantum physics's more famous solved problems. I have limited space in this chapter to delve too deeply into the problem-solving, so after I introduce the problems, I direct you to some Web sites where you can find more in-depth info about the ins and outs of each of these problems.

Finding Free Particles

Finding the free-particle wave function (and discussing wave packets) is an important aspect of quantum physics. A *free particle* is just that — one that doesn't feel any forces or isn't constrained by any boundaries. Yet free-particle wave functions have to be physical — they can't go to infinity, for example, which is why the idea of wave packets was introduced.

Check out `rugth30.phys.rug.nl/quantummechanics/potential.htm` for an example. This tutorial also has some QuickTime videos on wave functions.

Enclosing Particles in a Box

You can use the free-particle wave function and wave packets and take those concepts a step further to enclose particles in a box — that is, trap them in a potential well. This scenario is much more true to life, because all particles come up against some kind of force sooner or later.

Take a look at `rugth30.phys.rug.nl/quantummechanics/potential.htm`. This site, which includes videos, discusses particles in a box.

Grasping the Uncertainty Principle

Quantum physics students are no strangers to uncertainty. The inability to find exact answers is a common theme in quantum physics, and the Heisenberg Uncertainty Principle is one of the more famous quantum physics problems. This principle says that the better you know something's position, the less well you know its momentum, and vice versa.

Visit `plato.stanford.edu/entries/qt-uncertainty` for a good discussion of solving the uncertainty principle in both momentum/position and energy/time.

Eyeing the Dual Nature of Light and Matter

One of quantum physics's most famous results is explaining how matter/particles and light waves can share many of the same properties. This result was found by accident when particles going through slits exhibited wave-like properties in addition to particle-like properties.

Go to `dev.physicslab.org/Lessons.aspx` and click on "Modern/Atomic" and then "An Outline: Dual Nature of Light and Matter." This site has a good discussion of the problems and their solutions.

Solving for Quantum Harmonic Oscillators

Another problem that quantum physics is good at solving is the quantum harmonic oscillator. A *harmonic oscillator* undergoes periodic motion, like a spring, and a *quantum harmonic oscillator* is just a harmonic oscillator on the scale of atoms.

Take a look at `fermi.la.asu.edu/PHY531/hogreen/node7.html` for a nice treatment of the math involved with quantum harmonic oscillators.

Uncovering the Bohr Model of the Atom

Solving for the quantized orbitals of the hydrogen atom is one of quantum physics's most famous solved problems. No model until the quantized Bohr model really explained the light spectrum you get from hydrogen atoms.

Check out `csep10.phys.utk.edu/astr162/lect/light/bohr.html` for a more in-depth discussion of the Bohr model of the quantized atom.

Tunneling in Quantum Physics

Like wayward dogs and children, particles have ways of getting places they're not supposed to be — according to the rules of classical physics, at least. The fact that particles can tunnel to locations that their energy seems to prohibit is one of quantum physics's most surprising results. *Tunneling* refers to particles that are where they wouldn't be allowed classically because they didn't have enough energy — for example, a particle that appears on the other side of a high energy barrier.

Go to `www.physicspost.com/science-article-173.html` for a good, five-page treatment of tunneling.

Understanding Scattering Theory

Treating particles as wave functions that can interact is the foundation of *scattering theory.* This model has proven remarkably successful at predicting the results of scattering interactions between particles.

Visit `www.ph.ed.ac.uk/~gja/qp/qp10.pdf` for a good overview of the topic.

Deciphering the Photoelectric Effect

Albert Einstein's early explanation of the photoelectric effect has always been one of quantum physics's prized solutions. The problem was that when you shine light on metal, it frees electrons; however, when you increase the intensity of the light, it doesn't increase the energy of the freed electrons, just their number. Einstein's solution of the problem — positing a quantized work function that electrons had to overcome before they could be free — explained the experimental results.

Check out `www.walter-fendt.de/ph14e/photoeffect.htm` for the intricacies of the photoelectric effect along with the apparatus used to detect it.

Unraveling the Spin of Electrons

Electrons spin inherently, but it took some heavy-duty experimental apparatus to determine that fact. You can hardly tell that electrons spin, but spin adds another quantum number to an electron's state. That means that two electrons can seemingly occupy the same quantum state in terms of position and the like, but if they have different spins, they won't interfere with each other.

Take a look at `hyperphysics.phy-astr.gsu.edu/Hbase/spin.html` for a good discussion of electron spin, intrinsic angular momentum, and the Stern-Gerlach experiment that discovered it all.

Chapter 14

Ten Ways to Avoid Common Errors When Solving Problems

In This Chapter

- Remembering definitions
- Using the right formulas

No one ever said solving quantum physics problems was easy. In fact, most of the time, it's downright difficult. You can encounter plenty of stumbling blocks along the way. But if you're going to make mistakes, you should at least make sure those mistakes involve the hard stuff — not, say, forgetting that little cross-shaped dagger symbol when you're writing an adjoint.

The good news is that you don't have to fall victim to common mistakes. This chapter helps you avoid some of the most common errors people make when solving quantum physics problems. Keep them in mind as you work through your problems.

Translate between Kets and Wave Functions

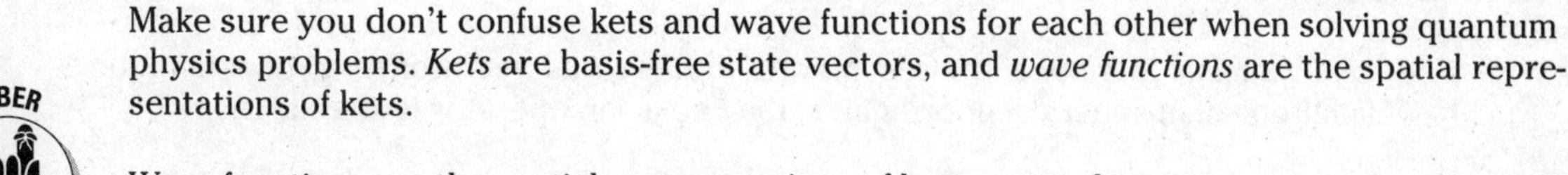

Make sure you don't confuse kets and wave functions for each other when solving quantum physics problems. *Kets* are basis-free state vectors, and *wave functions* are the spatial representations of kets.

Wave functions are the spatial representations of kets, so you have

$$\langle x|n\rangle = \psi_n(x)$$

Take the Complex Conjugate of Operators

When you reverse a bra and ket pair — for example, when you take them from one side of an equation to another — don't forget to take the complex conjugate of any operators. That is, $\langle\psi|A|\phi\rangle \neq \langle\phi|A|\psi\rangle$. Instead

$$\langle\psi|A^\dagger|\phi\rangle = \langle\phi|A|\psi\rangle$$

Take the Complex Conjugate of Wave Functions

When you *normalize* a wave function, you make sure that its square integrates to 1 over all space. When you're normalizing, bear in mind that the term $|\psi(x)|^2$ is made up of the wave function multiplied by its complex conjugate (where the imaginary parts change sign). Don't neglect to take the complex conjugate, as follows:

$$\langle x \rangle = \int_0^a x \left|\psi(x)\right|^2 dx$$

Include the Minus Sign in the Schrödinger Equation

When using the Schrödinger equation, you have to include the minus sign in front of the kinetic energy term. The momentum operator includes $\hbar/i$, and when you rationalize the fraction, multiplying the imaginary number by itself in the denominator gives you –1; therefore, you need the minus sign at the beginning to give you a positive kinetic energy.

Don't forget the minus sign in the momentum term of the Schrödinger equation — many people leave that out by mistake:

$$\frac{-\hbar^2}{2m}\nabla^2\psi(r) + V(r)\psi(r) = E\psi(r)$$

Include sin θ in the Laplacian in Spherical Coordinates

The *Laplacian operator* is the second-order differential operator that appears in the Hamiltonian. In spherical coordinates, the Laplacian operator looks like this:

$$\nabla^2 = \frac{1}{r}\frac{\partial^2 r}{\partial r^2} - \frac{1}{\hbar^2 r^2}\mathbf{L}^2$$

where $\mathbf{L}^2$ is the square of the orbital angular momentum. Never forget the sin θ term in the spherical coordinates version of the Laplacian:

$$\mathbf{L}^2 = -\hbar^2\left(\frac{1}{\sin\theta}\frac{\partial}{\partial\theta}\left(\sin\theta\frac{\partial}{\partial\theta}\right) + \frac{1}{\sin^2\theta}\frac{\partial^2}{\partial\phi^2}\right)$$

Remember that λ << 1 in Perturbation Hamiltonians

When you create a perturbation Hamiltonian, make sure that $\lambda << 1$ (that λ is much smaller than 1); inadvertently making a perturbation too large is easy. The perturbation Hamiltonian looks like this:

$$H|\psi_n\rangle = (H_0 + \lambda W)|\psi_n\rangle = E_n|\psi_n\rangle \quad (\lambda << 1)$$

For details on perturbation theory, see Chapter 10.

Don't Double Up on Integrals

Make sure you get the limits of integration right. It's easy to fall into the trap of doubling up — that is, ending up with twice the answer because you used twice the integration interval (one angle varies from 0 to 2π, and the other varies only from 0 to π to cover all space).

When you're integrating over space using spherical coordinates, don't forget that the second integral goes to π, not 2π:

$$\int_0^{2\pi} d\phi \int_0^{\pi} Y_{lm}^*(\theta,\phi) Y_{lm}(\theta,\phi) \sin\theta \, d\theta = 1$$

Use a Minus Sign for Antisymmetric Wave Functions under Particle Exchange

When you exchange particles in an antisymmetric wave function — that is, when you exchange one particle in a multi-particle wave function with another particle in the same wave function — don't forget to include a minus sign in the front:

$$P_{ij}\psi(\boldsymbol{r}_1, \boldsymbol{r}_2, \ldots, \boldsymbol{r}_i, \ldots, \boldsymbol{r}_j, \ldots, \boldsymbol{r}_N) = -\psi(\boldsymbol{r}_1, \boldsymbol{r}_2, \ldots, \boldsymbol{r}_i, \ldots, \boldsymbol{r}_j, \ldots, \boldsymbol{r}_N)$$

Remember What a Commutator Is

Commutators are important pieces to solving quantum physics problems, particularly when you want to move one operator past another one in an equation. As silly as it sounds, some people forget the definition of a commutator. Remember the following definition:

$$[A, B] = AB - BA$$

In other words, the commutator is the difference between using one operator first and then using the other operator first.

Take the Expectation Value When You Want Physical Measurements

When quantum physics looks for an actual physical measurement, it takes the expectation value. An *expectation value* is the physical value you expect, on average, of a particular measurement. (Refer to Chapter 1 for more info.) Don't forget about the expectation value when you're ready to get a physical measurement:

$$\text{Expectation value} = \langle \psi | A | \psi \rangle$$

Index

• Numerics •

• A •

• B •

• C •

• D •

• E •

• F •

• G •

• H •

• I •

• K •

• L •

• M •

• N •

• O •

• P •

• Q •

• R •

• S •

• T •

• U •

• V •

• W •

• Z •

Notes

Notes

Notes

Notes

Notes

Notes